FOOD-GRAINS OF INDIA

FOOD-GRAINS OF INDIA.

A. H. CHURCH

GYAN

Published by

Gyan Publishing House
5, Ansari Road
Daryaganj, New Delhi-110002
Phone: 011-47034999, 9811692060
E-mail: books@gyanbooks.com

Distribution Network
gyanbooks.com
India, USA, Canada, UK, Australia, France

ISBN: 978-81-212-8710-4 (PB)
First Published, 1886

2nd Impression 2023

Printed at: Gyan Press, Delhi.

FOOD-GRAINS OF INDIA
Author: A. H. Church

SOUTH KENSINGTON MUSEUM SCIENCE HANDBOOKS.

FOOD-GRAINS OF INDIA.

"National Foods" are illustrated by several small groups of specimens in the "Food Collection" now exhibited in the Bethnal Green Branch of the South Kensington Museum. In that Collection examples will be found of the chief Food-Grains of India described in this Handbook: Rice, Wheat, and many kinds of Pulse are represented by numerous specimens, the majority of which once formed part of the India Museum.

FOOD-GRAINS OF INDIA.

A. H. CHURCH,

M.A. OXON., F.C.S., F.I.C.

Professor of Chemistry in the Royal Academy of Arts, London ;
Sometime Professor of Chemistry in the Royal Agricultural College, Cirencester ;
Author of "Food, its Sources, Constituents, and Uses ;"
"The Laboratory Guide for Agricultural Students," etc., etc.

With Numerous Woodcuts.

Published for the Committee of Council on Education

BY

LONDON : CHAPMAN AND HALL,

LIMITED.

1886.

CONTENTS.

	PAGE
PREFACE ...	vii
LIST OF ILLUSTRATIONS...	ix
BIBLIOGRAPHICAL NOTES	xi

PART I.—THE CONSTITUENTS AND USES OF FOOD:

Water in Food	1
Salts or Mineral Matter	2
Albuminoids, Proteids, or Flesh-formers ...	2
Starch, Sugar, and Carbohydrates ...	4
Fibre, or Cellulose and Lignose	6
Oils and Fats	7
Food as Building and Reparative Material; as Warming Material or Fuel; as the Source of Energy	8
Nutrients and Food-Adjuncts	9
Classification of Food	10

PART II.—DIETARIES AND RATIONS:

A Mixed Diet	12
Diet and Work	13
The Nutrient Ratio: Nutrient Value	14
A Day's Ration	16
Tables for Calculating Rations	23
Examples of Rations	25

PART III.—THE CEREALS; CHARACTERISTICS AND CLASSIFICATION:

Millets	34
Maize	65
Rice ...	66
Wheat	90
Bamboo Rice	102

PART IV.—BUCKWHEAT AND ITS ALLIES; CHARACTERISTICS AND CLASSIFICATION:

PAGE

Amaranths 104
Quinoa 110
Buckwheat 114

PART V.—PULSE; CHARACTERISTICS, PREPARATION, AND CLASSIFICATION:

Lupines 123
Chick-peas 128
Vetches 131
Soy-beans 140
Haricots 147
Gram... 148
Pigeon-peas 169

PREFACE.

THE present Handbook has been prepared mainly with the object of furnishing to Indian officials and to students of Indian agriculture a compact account of the alimentary value of the chief Food-Grains of our Eastern Empire. A few notes on the cultivation of some of the crops have been incorporated with the work, while an endeavour has been made in two introductory chapters to show how a knowledge of the composition of the several Food-Grains may be utilised in the fixing of rations and the adjustment of dietaries.

The materials upon which I have drawn in compiling this volume were in great part gathered by Dr. J. Forbes Watson, who projected and, in some measure, carried forward towards completion an exhaustive treatise on Indian Foods. With that work in view, Dr. Forbes Watson not only collected much scattered information from a number of books, memoirs, and official documents, but caused many chemical analyses to be made; some of these were published in the "Journal of the Society of Arts" so long ago as 1857. These analyses refer almost exclusively to air-dried Food-Grains: they were performed with every care, and, in some directions, were extended so far as to include quantitative determinations of each of the ash-constituents of the products under examination. I have not found it possible, in a concise Handbook such as the present, to avail myself of all the data furnished by Dr. Forbes Watson's collections; but, on the other hand, I have supplied some missing facts and figures by means of inquiries and researches of my own. In the pages that follow the new analyses which I have myself made are indicated by the letter (C); the rest are Dr. Forbes Watson's. In not a few instances, however, the latter analyses have been controlled by fresh determinations of some of the constituents of the grains and seeds, and by the examination of additional and authenticated specimens.

Specimens of most of the grains and seeds described in this Handbook may be seen in the Museums of the Royal Gardens, Kew, and in the Bethnal Green Branch of the South Kensington Museum. The collection at Bethnal Green is particularly rich in specimens illustrating many out of the hundreds of varieties of Indian rice; the examples there presented of Indian wheat and Indian pulse are also numerous.

Should a second edition of this brief and incomplete Handbook be called for, I trust that I shall be able to fill in many of its gaps and to make it more exact. The Indian and Colonial Exhibition at South Kensington will doubtless furnish me with ample materials for extension and improvement. I venture to express a hope that I shall not be held responsible for the rendering of the Indian names introduced into my pages, for I have no first-hand knowledge of the subject, and, having been obliged to gather the names of places and products from a great variety of sources, neither accuracy nor uniformity of transliteration has been secured.

The chief works consulted in the preparation of this Handbook are recorded in the appended Bibliographical Notes; other books therein named will be found useful by persons desirous of further information concerning Indian Plants and Food Crops. But, after all, it is to the indefatigable labours of Dr. Forbes Watson that I am chiefly indebted, and it is to him that any merits to be found in this volume must be, for the most part, attributed.

May, 1886.

LIST OF ILLUSTRATIONS.

Figure	Name of Plant.	Page
1	Paspalum scrobiculatum	38
2	Panicum miliaceum	41
3	Panicum miliare...	45
4	Panicum frumentaceum	48
5	Panicum colonum	51
6, 7	Setaria italica	54
8, 9	Pennisetum typhoideum	56, 57
10	Coix Lachryma	61
11	Zea Maÿs	64
12, 13	Oryza sativa	67, 71
14	Saccharum officinarum	77
15, 16	Sorghum vulgare	81, 83
17, 18	Eleusine coracana	88
19	Amarantus paniculatus	106
20	Chenopodium Quinoa	111
21	Fagopyrum esculentum...	115
22	Arachis hypogæa	126
23	Cicer arietinum	129
24	Lathyrus sativus...	133
25	Lens esculenta	137
26	Glycine Soja	141
27	Canavalia ensiformis	145
28	Phaseolus Mungo	149
29	Phaseolus aconitifolius	153
30	Vigna Catiang	157
31	Dolichos Lablab...	160
32, 33	Dolichos biflorus	163, 165
34, 35	Cajanus indicus	168, 171

BIBLIOGRAPHICAL NOTES.

BIRDWOOD, SIR G. "Economic Products of Bombay." Bombay, 1862.

CHURCH, A. H. "Food, its Sources, Constituents, and Uses." London, 1876.

COOKE, DR. M. C. "Report on the Oil Seeds in the India Museum." London, 1876.

COTTON, J. S. "Moral and Material Progress of India, 1882–83." (Blue Book.) London, 1885.

DRURY, COL. H. "Useful Plants of India." London, 1873.

DUTHIE, J. F., and FULLER, J. B. "Field and Garden Crops of the North-West Provinces and Oudh." Parts i. ii. Roorkee, 1882–83.

HOOKER, SIR J. D. "Flora of British India." London, 1875–85.

HOOKER, SIR J. D., and BENTHAM, G. "Genera Plantarum." London, 1867–80.

HUNTER, DR. W. W. "Imperial Gazetteer of India." Second Edition. London, 1886.

KÖNIG, DR. J. "Chemie der Menschlichen Nahrungs- und Genussmittel." Berlin, 1879–80.

LYON, J. B. "Food Tables." Bombay, 1877.

MEINERT, DR. C. A. "Armee- und Volksernährung." Berlin, 1880.

POWELL, BADEN H. "Economic Products of the Punjab." Roorkee, 1868.

ROXBURGH, DR. W. "Flora Indica." Calcutta, 1870.

WATSON, DR. J. FORBES. MS. Collections; Original Memoranda and Analyses; and a Paper in the "Society of Arts Journal," November 27, 1857.

WATSON, DR. J. FORBES. "Indian Products; Industrial Survey of India." Part ii. London, 1872.

WATT, G. "Economic Products of India. Part vi. Foods." Calcutta, 1883.

FOOD-GRAINS OF INDIA.

PART I.

THE CONSTITUENTS AND USES OF FOOD.

Water.—This important compound, a combination of 2 parts by weight or 2 proportions of the element hydrogen, with 16 parts by weight or 1 proportion of the element oxygen, forms more than two-thirds by weight of the adult human body, and 80 per cent. of that of the infant. It enters largely into the very structure of the tissues, being, in fact, an actual food, and it acts as a carrier of food into and through the system. As a general rule our total daily meat and drink should contain about four-fifths of their entire weight of water, and in hot countries and seasons a good deal more. Milk, the natural and model food for infants, contains a proportion of water which is even larger than this, namely, 88 per cent. The average amount of water in the majority of ripe and air-dried Indian cereals, such as wheat, rice, and millets, and in the various kinds of Indian pulse, is one-eighth of their entire weight, or 12·5 per cent. This proportion of water varies, however, a little with different kinds of grain: in temperate climates it stands at a higher level than in the tropics and in semi-tropical countries, perhaps amounting to 14½ per cent. In leaves the proportion of water is much greater than in grains and pulse; in succulent stems, roots, and fruits, it reaches its maximum—in many kinds of melons, for example, 90 to 95 parts of water in the hundred is no unusual proportion. In the muscular tissue or pure lean of meat the water constitutes three-fourths of the whole.

Mineral Matter or Salts.—Although nearly all the constituents of food are capable of being burnt, and are therefore spoken of as combustible compounds, yet water and certain necessary saline matters are not the less nutritious because they cannot be burnt. The human body contains in every tissue and secretion some mineral matter; in the bones and teeth a great deal. Phosphates, carbonates, sulphates, and chlorides are the chief among the salts of the body. Lime, potash, soda, and magnesia are the metallic bases of these salts. Sometimes, however, these bases are united, not with the mineral acids (phosphoric, sulphuric, etc.), but with the acids called organic, which invariably contain carbon. The mineral matters are generally supplied in sufficient quantity by an ordinary vegetable diet—the "ash" which figures in all analyses representing roughly the amount of fixed or mineral matter present. Rice is perhaps the only important food in which these compounds are deficient. Common salt, however, has everywhere to be added to food as its two constituents, chlorine and sodium, are present in very small proportion in vegetable products, save in those which grow by the sea-shore or in the sea. Iron and silica, of which minute quantities are essential to the nourishment of the body, are furnished in ample quantity by all kinds of vegetable foods. In the analyses recorded in the following pages, some figures are given representing the percentages of potash and phosphoric acid in several food-grains; it will be noticed that those grains or seeds from which the husk has been removed are poorer in ash than those which have not been so "cleaned."

Albuminoids, Proteids, or Flesh-formers.—From a colourless substance, which is the chief constituent of the albumen or white of egg, a whole group of related organic compounds containing nitrogen has been named "albuminoids." From the variety of forms in which they appear, and the numerous changes they undergo, the term "proteids" has also been applied to them; while from a function in the body which they alone are competent to perform, they have been designated "flesh-formers." The difficulty of separating the albuminoids in a pure and unaltered state has prevented chemists from unravelling their chemical constitution, which is, however, known to be exceedingly complex. All the albuminoids are either naturally soluble in water, or may be made so by the action of a fixed alkali, the so-called albuminates being thus formed. Their chief element, characterising by its presence the whole group, is nitrogen—the element which constitutes seventy-nine parts by measure of common air, which is present in nitre, nitric acid, and ammonia, and which is generally so much more abundant in animal than in vegetable tissues. The percentage of nitrogen in the different albuminoids appears to range between 13 and 18½, but there are doubts as to the purity of the materials which, when submitted to analysis, have yielded these numbers. The range of variation in the percentage composition of the members of the whole group is shown in this table:

Carbon	50·6 to 54·3	per cent.
Hydrogen	6·7 „ 7·3	„
Nitrogen	12·8 „ 18·5	„
Sulphur	·4 „ 1·7	„
Oxygen	20·6 „ 27·0	„

The amount of albuminoids present in any food is usually ascertained by determining the amount of nitrogen existing in it, and then multiplying the nitrogen found by the factor 6·3. In this procedure two assumptions are made: one, that all albuminoids contain a mean percentage of 15·87 of nitrogen, for $6{\cdot}3 \times 15{\cdot}87 = 99{\cdot}981$, practically 100; the other assumption is that no other nitrogen compounds are present. The former assumption may pass, the latter needs a word of explanation. Vegetable and animal foods do contain nitrogen in non-albuminoid forms; for instance, more than half the nitrogen of lettuces, water-cress, and spinach exists in the form of nitrates, which are wholly useless as flesh-formers. Then, too, there are a number of other nitrogen-compounds in organic food products which are non-albuminoid. Amongst these may be named the ossein and cartilagin of bone and cartilage, and the amides of beet-root, potato tubers, and unripe pulse. But it so happens, that the ripe grains and seeds which are discussed in the present Handbook generally contain but small quantities of nitrogen in forms other than those of the albuminoids, so that we shall not be led into serious error by ignoring their existence. In some of these food-grains, however, small quantities of the compounds called *peptones* have been found, but these are known to fulfil the same nutritive functions as albuminoids, and indeed are the forms in which the latter after digestion are taken up into the circulatory system of the human body; they are diffusible through membranes which the original albuminoids are not. Further explanations as to the nitrogen-compounds will be found on pp. 32 and 118.

Starch, Sugar, and Carbohydrates in general.—The carbohydrates are so called because they may be regarded as containing hydrogen and oxygen in the proportion in which they form water, associated with carbon. The most important member of the group is starch. This is the chief constituent of all dry food-grains, except those which contain very much oil. Starch occurs mainly in the form of granules, which differ much in different grains both in size and shape, and, it may be added, in their rate of digestion. Uncooked maize-starch and oat-starch are very quickly turned into sugar by the digestive ferment of the saliva or by the diastase of the pancreas; wheat-starch requires a much longer time, and potato-starch is still more obstinate; when thoroughly cooked all starches require nearly the same time for salivary and pancreatic digestion. The products of this change are dextrin, maltose, and a little grape-sugar. Dextrin may also be formed by gently roasting starch and in many other ways; it is more easily digested than starch, but has the same percentage composition, and probably the same nutritive power. It occurs abundantly in parched grain and in the crust of bread. Several varieties of dextrin exist.

The chemical expression for starch and dextrin is $n\ (C_6H_{10}O_5)$, a formula which corresponds to the following percentages:—Carbon 44·4, Hydrogen 6·2, Oxygen 49·4.

The term "sugar" is commonly applied to a number of different, although related substances, possessing a more or less marked sweet taste. These substances are arranged by chemists in three groups, of which cane-sugar, grape-sugar, and manna-sugar may be taken as the respective types.

Group i.—The chemical formula for the sugars of the first group is $C_{12}H_{22}O_{11}$, an expression which corresponds to the following percentages:—Carbon 42·1, Hydrogen 6·4, Oxygen 51·5.

a. Cane-sugar occurs in the sugar-cane (*Saccharum officinarum*) and in many other grasses; in some palms, as Borassus flabelliformis, Caryota urens, and Arenga saccharifera; in beet-root (*Beta vulgaris*); in the carrot (*Daucus Carota*); the parsnip (*Pastinaca sativa*); in some maples, as Acer saccharinum and three other North American species; and in the nectaries of many flowers.

b. Milk-sugar has been found in the latex of Achras Sapota.

c. Melitose occurs in the manna exuding from many species of Eucalyptus.

d. Melezitose is found in the "manna" which exudes from the young twigs of the larch; it is the chief constituent of "Lahore Manna," from the Camelthorn, Alhagi maurorum (*Hedysarum alkagi*, L.), a shrub growing in Persia and Afghanistan.

e. Mycose occurs in "trehala manna," known in Persia as "nest-sugar;" it is also a constituent of several fungi.

f. Maltose occurs naturally to a small extent in many cereal grains, and more abundantly in malt.

Group ii.—The chemical formula for the sugars of the second group is $C_{12}H_6O_6$, an expression which corresponds to the following percentages:—Carbon 40·0, Hydrogen 6·7, Oxygen 53·3.

a. Dextrose, or grape-sugar, is the hard crystalline sugar so abundant in raisins; it is found in ripe cherries, and, associated with another glucose, in honey and in many immature fruits.

It is largely made from starch and cellulose; seventeen parts of it are equal in sweetening power to ten of cane-sugar.

b. Lævulose, or fruit-sugar, accompanies dextrose in honey and in many acid fruits; sun-dried Mahúa flowers (*Bassia latifolia*) contain about half their weight of "invert-sugar," a mixture of equal parts of lævulose and dextrose.

c. Inosite is found in the leaves of some plants, and in the unripe pods of several kinds of pulse.

Group iii.—There are at least five varieties of manna-sugar:

a. Mannite ($C_6H_{14}O_6$), in celery roots, olives, and some algæ and fungi; and largely in ordinary manna, the exudation of certain ash-trees, as Fraxinus Ornus and F. integrifolia.

b. Dulcite ($C_6H_{14}O_6$), occurs in Madagascar manna, and in the sap of several plants, as Rhinanthus crista-galli, Melampyrum nemorosum, and Euonymus europæus.

c. Sorbite ($C_6H_{14}O_6$), is found in mountain-ash berries.

d. Pinite ($C_6H_{12}O_5$), occurs in the sap of Pinus Lambertiana.

e. Quercite ($C_6H_{12}O_5$), is found in acorns.

Gum, Mucilage and Pectose.—The various gummy and mucilaginous substances occurring in plants either dissolve in water or swell up therein to a jelly. They invariably leave a good deal of ash when burnt, and are, in fact, lime, magnesia, and potash salts of certain weak organic acids. Gum arabic, gum tragacanth, and linseed mucilage may be cited as examples of this group of nutritive substances. To these may be added pectose and its derivatives—substances which occur abundantly in fleshy fruits, and which are nearly related to mucilage. The exact values as food of the members of this group are not known, but they may safely be assigned a position somewhat inferior to starch and sugar.

Fibre, Cellulose, etc.—Man does not appear to have the power of digesting the fully-formed cellular tissues of plants. The basis of these is a substance called cellulose, but it is probable that there are several slightly differing compounds included under this term; the percentage composition of cellulose is that of starch. Fibre is not an homogeneous and uniform substance, but a varying mixture of cellulose with a complex body or bodies, called lignose or lignin, related to the aromatic series. In the following pages we use the term "fibre" as a convenient designation for that part of the food-grain which is neither albuminoid, nor oily, nor starchy, which resists the solvent action of boiling alkaline and acid solutions, and which is unaffected by alcohol and by ether. It contains both cellulose and lignose (sometimes cutin), but however variable and mixed its composition it may be safely affirmed to be wholly useless as food. In actual foods it is indeed often worse than useless, locking up in an inaccessible condition a part of the albuminoids, the oil, and the mineral matter of the seed or grain.

*Oils and Fats.**—The vegetable oils and fats, while insoluble in water hot or cold, are readily dissolved by liquids which are themselves not miscible with water, such as benzene, petroleum-oils, and ether. They are distinguished from the volatile oils by their leaving a permanent translucent trace on paper, and by their fixity at ordinary temperatures. Very few of the seeds discussed in the present Handbook contain so much oil that it can be extracted by simple pressure; the pea-nut (p. 127) is an exception. The fixed vegetable oils and fats consist of certain chemical compounds called glycerides, associated with small quantities of substances known as "fatty acids." The glycerides themselves may be regarded as glycerin, plus one or more fatty acids, and minus water. While starch and sugar contain hydrogen and oxygen in the exact proportions to form water, oils contain an excess of hydrogen. From this cause, and from their greater richness in carbon, a given weight of oil will give out more heat, or set free more energy when burning, than the same weight of starch or sugar. The following table shows the percentage composition of the oils

---

* These are sometimes called "hydrocarbons" by writers on dietetics; this is incorrect and misleading. True hydrocarbons contain hydrogen and carbon only, and are absolutely indigestible and useless as food.

extracted by means of ether or of benzene from two cereals and two pulses.

	Carbon.	Hydrogen.	Oxygen.
Wheat Oil	75·97	11·33	12·70
Rice Oil	76·28	12·40	11·32
Pease Oil...	74·37	11·32	14·31
Bean Oil	75·70	11·86	12·44

If we accept Carbon 76, Hydrogen 11·5, and Oxygen 12·5, as representing the average composition of the oils in question, we shall probably not be far wrong.

It should be noted that many oils expressed from seeds or fruits are added to the other staple articles of Indian diet. Amongst these may be cited the oils and fats expressed from the coco-nut, and the fruits from other palms; from safflower, niger, and sunflower seeds amongst the composites; from mustard, rape, and gold of pleasure seeds amongst the crucifers; from linseed, and from sesame seed. These oils do not differ appreciably from the oils extracted from cereals and pulse, so far as their food value is concerned. On the other hand, animal oils and fats, such as clarified butter or milk-fat, contain no free fatty acids, consisting wholly of the glycerine compounds of stearic, palmitic, oleic, butyric, and other allied acids.

Uses of Food.—A very little reflection will show that the office of food in the body is threefold. The daily and continuous material losses of the body must be made good; this is accomplished by food as a building and reparative material. Secondly, the warmth of the body must be maintained by food as a warming material or fuel. Thirdly, food must constantly supply the amount of energy necessary for the performance of the external and internal work of the body. The human body has often been likened to a locomotive steam-engine. With some important differences there are striking analogies between these structures. In the one case there is food, in the other coal; in the one case there is an incessant and considerable consumption of material which has become part and parcel of the various bodily organs, in the other the losses of the structure or mechanism itself are neither essential nor large. In both structures, however, carbon and hydrogen are burnt, yielding the same products; in both water is heated and vaporised; in both stored up, or potential energy is converted into actual energy and heat. From both structures carbonic acid and steam escape; from both structures incombustible or unused waste is thrown out. From the points of comparison and of contrast just indicated, it is evident that food has to perform one function which is not demanded of engine-coal, namely the repair and even the building of the structure or apparatus in which it is consumed. Moreover, there are but a few substances which can be used as food in the human body, while the steam-engine may consume coal or coke or petroleum—substances which, to the organs of man, are absolutely useless.

Nutrients and Food-Adjuncts.—We are now in a position to classify the various useful constituents of human food according to the functions they perform and their chemical constitution. But there are many substances consumed by man which cannot be regarded as actually nutritive, but which, by a universal instinct, are associated with his daily food. We refer to the group of vegetable substances which includes spices, condiments, tea, coffee, etc. While the essential foods may be designated "nutrients," these may be called "food-adjuncts." They are useful in stimulating the flow of the various digestive secretions of the alimentary tract, and in making food of palatable and varied flavour. Many of them have also a marked and favourable influence on the nervous system.

It will be seen, from what has been already stated, that the nutrients fall into two divisions, incombustible and combustible. To the former belong water and salts: to the latter, certain compounds containing carbon, hydrogen, and oxygen, and another group of compounds in which nitrogen also exists. All the combustible nutrients serve, or may serve, as sources of heat and energy in the system, but to those containing nitrogen a further office belongs—

that of building up and renewing the blood and tissues of the body. The "potential" energy laid up in each of the combustible nutrients is capable of being set free as actual or kinetic energy, when the nutrient, after digestion, absorption, and assimilation, is oxidized in the organism by means of the oxygen taken into the lungs. But the nitrogenous nutrients alone are capable of transformation into the very substance of the nitrogenous tissues themselves.

CHEMICAL AND PHYSIOLOGICAL CLASSIFICATION OF FOOD.

CLASS I.—NUTRIENTS.

DIVISION I.—INCOMBUSTIBLE COMPOUNDS.

i. WATER :—The carrier of nutritive materials and waste products ; forms an essential part of all tissues.

ii. SALTS OR MINERAL MATTER :—Such as *common salt* and *phosphate of lime*, which serve to effect solution and change, and to build up certain tissues.

DIVISION II.—COMBUSTIBLE COMPOUNDS.

iii. CARBON-COMPOUNDS, such as *starch*, *sugar*, and *oil*, which serve to keep up the heat, and do the work of the body by the discharge of their potential energy during oxidation in the organism. The fat of the body is formed in part from the fat or oil in the food ; in part from the starch and sugar. Gum, mucilage, pectose, etc., belong here.

iv. NITROGEN-COMPOUNDS, such as *albumen*, *fibrin*, and *legumin*, the chief formative and reparative compounds of food ; they may also yield fat, and by their oxidation set free heat and actual energy. Hereafter we shall call them "albuminoids."

CLASS II.—FOOD-ADJUNCTS.

i. ALCOHOL, as contained in *fermented* and *distilled liquors.*

ii. VOLATILE OR ESSENTIAL OILS and other odorous and aromatic compounds, as contained in *condiments* like garlic, turmeric, and pepper, and in *spices* as ginger and cloves.

iii. ACIDS, as *citric acid* in lemons, *acetic acid* in pickles.

iv. ALKALOIDS, as *caffeine* in coffee and tea, and *nicotine* in tobacco.

PART II.

DIETARIES AND RATIONS.

A Mixed Diet.—Food in order to fulfil the various offices described in the preceding pages must not contain one set of nutrients only, but all. Moreover it must contain the several nutrients in a certain ratio or proportion as well as to a certain amount. Absolute exactness of ratio or of amount is not required, as a small excess of one or more nutrients may remain unused, or may, in some cases, do the work of others. Thus an excess of water or of common salt is excreted without doing harm, and, within certain limits, albuminoids may do the work of oil or starch. Indeed, the human body may be sustained for a time upon water and albuminoids alone, although in this instance, the extra strain thrown upon those secretions and viscera which have to digest and to eliminate the excess of nitrogenous matters, could not be maintained for long without resulting in disease. From these and other considerations we may learn the necessity of a mixed diet by means of which each secretion and each organ of the body shall be properly utilised and perform its proper share in the work of nutrition. There must be starch, to be dissolved and saccharified by the ptyalin of the saliva and by the diastatic ferment of the pancreatic secretion; there must be oil to be emulsified and saponified and decomposed by the bile and the pancreatic juice; there must be albuminoids to be peptonised by the pepsin of the gastric juice, and by the trypsin of the pancreas. And the mixed food as a whole must be palatable and digestible, and it must occupy a sufficient but not excessive space.

Diet and Work.—It has been calculated from the results of numerous analyses of common and efficient dietaries, and by means of direct experiments with artificially constructed rations, that an adult man weighing 140 to 150 lbs. requires per day about

Albuminoids	4·16 oz.
Oil or Fat	3·12 „
Starch	14·29 „

These are the figures used by Forster and Voit: they correspond to about 282 grains of nitrogen and 5060 grains of carbon. For reasons connected with the smaller body-weight of native Indians, with the climate, and with the special vegetable diet of the East, we have reduced and modified these amounts to some extent in the examples of standard rations given later on.

It may be interesting and useful to state here that the amount of actual energy expressed in *calories* or heat-units which the above standard diet is capable of setting free is calculated to be:

4·16	ozs.	Albuminoids	give	503,034
3·12	„	Fat or Oil	„	801,699
14·29	„	Starch	„	1,534,600
				2,839,333 Total Calories.

These figures are approximations only, but they serve to show that there is a large difference between the amounts of energy expressed as heat which the three nutrients above-named can set free, fat being more energetic than albuminoids, and albuminoids more energetic than starch. The most recent experiments lead one to the conviction that 1 part of vegetable oil or fat is practically equal to 2·3 parts of starch: we accordingly use 2·3 as the "starch-equivalent" of one part of oil.

The heat which could be set free by the oxidation of the nutrients in a day's ration is partly transformed into its mechanical equivalent of work. The mutual convertibility of and quantitative relation between work and heat is generally expressed by saying that 1 lb. of water is heated 1° Fahr. by the amount of actual energy which is sufficient to raise 772 lbs. to a height of 1 foot. The amount of work done in and by the human body demands a very considerable proportion of the actual energy derived from the daily ration, the remainder being used to warm the body itself, the air inspired, the food consumed, the water vaporised from the skin and lungs, and the waste excreted. Of the energy employed in or applicable to mechanical work inside the body or external to it, a very large proportion is consumed in the work done by the heart and in other automatic internal labour : this has been estimated at 260 "foot-tons"—that is, it is equivalent to the work which a man does who raises 260 tons to the height of 1 foot. The average amount of external work, expressed in the same way, which an ordinary European labourer performs may be set down as equal to 300 foot-tons ; a country postman walking a daily round of 20 miles does work equal to 353 foot-tons ; it has been calculated that the pedestrian Weston, travelling 50 miles a day, did no less than 793 foot-tons of work per diem. Generally speaking, one-fifth of the energy derivable from the food is available for external work, and 2 foot-tons per 1 lb. of the body-weight is a fair day's work.

It appears that a larger proportion of nitrogen to carbon (that is, of albuminoids to starch and oil) is naturally demanded as the stress and amount of labour required becomes greater. This increase was formerly attributed to the greater waste assumed to occur in the nitrogenous muscular tissue during increased exertion. It is more probable that it is due to the comparative ease with which albuminoids, especially those of meat, are assimilated during strenuous labour, yielding urea as an excretory product on the one hand, and on the other fat, which serves as a muscular fuel. Another part of the nitrogenous food goes to augment the number of the red corpuscles of the blood. These corpuscles have a specially increased oxygen-carrying office to perform during violent or strenuous labour.

Nutrient-Ratio and Nutrient-Value.—The term "nutrient-ratio" has been used in this Handbook to designate the proportion of albuminoids to starch—including with the starch the starch-equivalent of any oil or fat present in the food. The expression "nutrient-value" refers to the sum total of the albuminoids, the starch, and the starch-equivalent of the oil. A complete nutrient-ratio is of course much more complex than the ratio which the albuminoids bear to the starch. It will include the water and the saline or mineral matter as well as the albuminoids, the oil, and the group of the carbohydrates. In fact, a further distinction would have to be made as to the several constituents of the dietary. For example, the proportions of potash, phosphoric acid, lime, magnesia, and especially of common salt, comprised in the "mineral matter" would have to be defined. The carbohydrates, too, must not wholly consist of sugar, although to starch, as a sole constituent, there is no objection. But, in practice, there is no necessity for such extreme precision, provided only that a *sufficient* supply of necessary nutriment be given. This end will be accomplished if due regard be paid, in framing a dietary, to the general principles laid down in this part of the present Handbook. The due admixture of food-grains belonging to different classes will secure the necessary mineral ingredients, provided an allowance of at least 230 grains* of common salt be added to the daily ration of a man having a body-weight of 105 lbs. The fat or oil, added to, or naturally present in the food-staples, should stand to the albuminoids in a proportion of not less than 1 to 2, and to the starch of 1 to 9—at least these proportions are found to give satisfactory results where moderate work only is demanded, although a further increase of the oil or fat, and a corresponding diminution of the amount of starch, may often be made with advantage. European model or standard dietaries generally contain a higher proportion of oil or fat, mainly owing to the very large use therein made of animal foods. For instance, it has been shown that a first-rate average daily diet for a European adult performing moderate labour may be advantageously constituted, in 100 parts, of

Albuminoids	3·9	with	Water	81·5
Starch, Sugar, etc.	10·6		Common Salt	0·7
Oil or Fat	3·0		Phosphates, Potash Salts, etc.	0·3

* The amount of salt consumed per head per day in the Bombay Presidency just exceeds ½ ounce; in Sind it is calculated to be little more than ¼ ounce.

It will be seen that the above percentages correspond to a ratio between the oil and the albuminoids of 1 : 1·3, and between the oil and the starch of 1 : 3·5—the "nutrient-ratio" being about 1 : 4·5.

Of the proportions to be introduced into a ration of a very important group of food-adjuncts, namely, the spices, condiments, and flavourers, no quantitative statement can be safely made. Examples of the amounts actually allowed per head per day have been published in several official Indian reports. One of these examples may be cited from the *Indian Economist*, of 15th October, 1870. Surgeon S. B. Partridge gives as materials for "curry stuff," ½ a drachm of garlic, ½ a drachm of chillies, and the same quantity of mustard; of black pepper, 1½ drachms; coriander fruits, 2 drachms; turmeric, 4 drachms; and tamarinds, 8 drachms. Four or five ounces of fresh vegetables, such as onions, potatoes, and melons, are extremely useful as supplying some of the constituents in which a rice or millet and pulse diet is deficient, and in adding variety, both of flavour and of texture or consistency to food, which otherwise might be monotonous or unpalatable. In fact, the dilution of such dry and concentrated food staples as rice, millet, and pulse, with succulent vegetables containing much water in intimate association with soft and pulpy cellulose, performs the very useful office of securing more complete digestion and assimilation of the starch, albuminoids, and oil.

From what has been said in the preceding paragraphs it is evident that the terms "nutrient-ratio," and "nutrient-value," as employed in this Handbook, have a somewhat restricted meaning. Nevertheless, these terms are extremely useful, for they serve to express the two chief or fundamental facts upon which rational dietaries must be constructed.

A Day's Ration.—We have just seen how, from the chemical analysis of actual dietaries in successful use, and from the results in practice of theoretically-constructed dietaries, several normal or standard rations have been devised in which defined quantities and proportions of albuminoids, fat, and starch are associated. Besides these ingredients it is necessary that a small but adequate amount of phosphoric acid and of potash should be present, while common salt is, of course, invariably added. No special provision for the supply of the other mineral or ash-constituents of human food need generally be made, as the lime, magnesia, iron, and sulphuric acid (or sulphur) required are furnished in abundance by most varieties of food, such, for instance, as wheat, buckwheat, and the millets. With rice, however, the case is different. This grain contains very little mineral matter, yielding on complete burning only about half a grain of ash for each hundred grains of cleaned rice taken for the purpose of the experiment. Fortunately, fresh fruits, green vegetables, and especially the different kinds of pulse, which are frequently used to supplement the other deficiencies of rice, supplement also its deficiency in mineral matter. Thus 1 ounce of pigeon-peas (*Cajanus indicus*) contains as much phosphoric acid and nearly twice as much potash as 3½ ounces of rice. So likewise, in adding pulse to rice in order to furnish the due proportion of albuminoids or "flesh-formers," we introduce also sufficient supplies of lime, magnesia, etc.

We may then leave out of consideration the mineral matter and the common salt, and confine our attention to the three main constituents of food, namely, the albuminoids, the oil, and the starch. Reference should be made to pp. 2, 4, 6, and 7 for explanations of the meanings attached to these three terms, especially as to the inclusion, under the name "starch," of sugar, dextrin, pectose, and other allied compounds, and also as to the expression "starch-equivalent of fat or oil;" this we assume to be 2·3, so that we may convert (for purposes of calculation) the oil of a ration into its starch-equivalent by multiplying it by this coefficient.

Although the various causes which combine to render a mixed diet not only desirable but necessary, if robust health is to be maintained, have been already described, it will serve to give special significance to the general argument on this point if we illustrate it here by a few selected examples. Scarcely any single article of food (and we confine ourselves in this Handbook to the food-staples of the vegetable kingdom) is complete and properly adjusted, even from a chemical point of view, much less in its combined chemical and physiological aspect. Even human milk, as a food for adults, is far from perfect. Its nutrient-ratio is 1 : 7, when 1 : 5 is required; and it contains no starch, although two special ferments are provided in the digestive secretions (saliva and pancreatic juice) for the purpose of dealing with this important carbohydrate. But when we examine the nutrient-ratios of the food-grains which are discussed in the following pages, we shall discover how very rarely any single food-staple presents the normal nutrient-ratio of a standard diet: in the few cases in which such a standard is approached, another defect becomes evident on closer inspection, namely, a deficiency in oil or fat. Take a few cases. Out of the forty or more important grains or seeds, four or five only approach sufficiently near to the standard nutrient-ratio of 1 : 5 to form satisfactory aliments when used alone for any length of time. These are:

Buckwheat (husked)	with the ratio		1 : 4·7
Wheat (Indian, whole, average)	,,	,,	1 : 5·2
Pea-nuts (seeds, whole)	,,	,,	1 : 5·2
Amaranths (seeds, whole)	,,	,,	1 : 5·3

And even amongst these four foods, one, namely wheat, would approach nearer perfection if it contained more oil, while pea-nuts would be more useful if they contained much less. If we take other instances, we shall find very extreme divergences from our standard ratio:

Ragí (Eleusine coracana),	with the ratio		1 : 13·0
Rice (cleaned)	,,	,,	1 : 10·8
Maize (whole)	,,	,,	1 : 8·3
Chick-peas	,,	,,	1 : 3·3
Pigeon-peas	,,	,,	1 : 3·0
Soy-beans	,,	,,	1 : 2·0

It is obvious from these figures that any one consuming sufficient ragí, rice, or maize to provide himself with the required amount, say 3 ounces per diem, of albuminoids, would have to consume along with these 3 ounces an immense excess of starch over that really wanted. In the case of the three cereals above named, the quantities in ounces of starch taken compared with those required would be as follow:

	Taken.	Wanted.	Excess.
Ragi	39.0 ...	15·8 ...	23·2
Rice	30·7 ...	15·8 ..	14·9
Maize	24·9 ...	15·8 ...	9·1

or, on the average, about one pound too much, daily, of starch. Of course a pulse diet presents exactly the converse defect, a diet of pigeon-peas, for example, furnishing less than 9 ounces of starch to 3 ounces of albuminoids. But it is clear that the defect of cereals as to the nutrient-ratio is precisely of that kind which can be neutralised by the due admixture of pulse; to that adjustment we now address ourselves.

However, several circumstances must be taken into consideration, in adjusting, on the above lines, the quantities and proportions of the several constituents of a day's ration. The body-weight and the amount of work to be done are the chief of these, but climate, age, peculiarities of food-materials available, and national or local tastes and prejudices, must not be left out of view. In hot countries less oily matter is needed and more water; the young require more fat and mineral matter than adults; condiments, spices, and other flavourers must be varied in kind or altered in quantity in different places. But, after all, the two conditions which it is specially necessary to discuss are those of body-weight and work. Most European standard dietaries have been constructed for the sustenance of individuals weighing from 140 to 154 lbs., and the work these individuals have been supposed able to perform is nearly proportional to such weight. If, then, we assume an amount of daily work performed by an average European labourer, weighing 150 lbs., can be expressible in this form—that he lifts 300 tons one foot—we shall find that natives of India weighing 115 lbs. can and do perform an amount of daily work equal to 230 foot-tons; and that if their weight be 105 lbs. only, they can and do accomplish work equal to 215 foot-tons. These amounts are strictly proportional to the lesser body-weights named, and are often exceeded in actual practice, though much less is demanded in Indian Government relief works.

We shall assume, in all our calculated dietaries and rations, that the body-weight is 105 lbs. only; it will be easy to add one-tenth to the calculated quantities given if the individuals to be fed weigh 115 lbs.; and we shall propose three dietaries: one (A) for bare sustenance; one (B) for moderate work; and one (C) for hard work. We have touched upon the principles underlying such dietaries in a previous section; into the various controversies as to the relations of nitrogenous and carbonaceous nutrients to work, this is not the place to enter; but for details of the most recent theories on this subject we may refer our readers to the work by Dr. C. A. Meinert, named in our "Bibliographical Notes," p. xi.

Here are the amounts needed daily for the three dietaries above indicated, expressed in avoirdupois ounces and decimals of an ounce.

Ration.	Albuminoids.	Oil.	Starch, etc.	Starch-equivalent.	Nutrient-ratio.
A	2·123	0·752	7·520	9·250	1 : 4·34
B	2·954	1·412	12·531	15·779	1 : 5·34
C	3·635	2·506	11·190	16·954	1 : 4·66

The attempt to realise these exact figures in actual practice would be useless; but there are many combinations of food materials which may be made to approach them with quite sufficient accuracy. Two guiding principles should not be lost sight of; namely, to keep down the pulse-constituent to an amount not exceeding 7 ounces per diem, and, if possible, not more than 5 ounces, and to ensure the presence of a sufficiency of albuminoids by increasing the cereal constituents of the ration, even if in so doing the quantity of starch required be thereby raised above the necessary amount. Indeed, in the case of a bare-sustenance ration, or where little or no labour is demanded from the persons fed, it has been proved beyond doubt that the starch may be largely increased and the albuminoids largely diminished without deranging the general health or lowering the body-weight. Thus a dietary in which the nutrient-ratio is 1 : 7 or even 1 : 8 has been adopted, in particular cases, with satisfactory results. It may be remarked that the total weight of pulse, cereal and added oil will vary, according to the richness of the pulse and cereal employed, so that in ration B, for instance, it may amount to no more than 20 ounces or it may rise to 22. No allowance has been made for the slight increase in the quantities of nutrients caused by the additions of fresh vegetables or of fruits to the dietaries, but it must be remembered that most of these products contain very large amounts of water, often 90 per cent., or even more. No notice is here taken of the question of condiments, nor of the desirability of varying from time to time the materials of the diet as far as practicable.

The two tables which follow represent in ounces and decimals of an ounce the amounts of albuminoids and of total starch (that is, of starch, plus the oil present converted into its starch-equivalent) in the stated numbers of ounces of various kinds of pulse and of cereal-grains. In constructing actual rations from these tables we must recollect that for dietary A, about 12 to 14 ounces of food-stuffs are needed; for dietary B, about 20 to 23 ounces; and for dietary C, about 21 to 25 ounces. The working of the tables may be gathered from the examples which follow them; of course approximations only can be reached. Reference must be made to the fuller analyses given under the heads of the several pulses, cereals, etc., when it is necessary to learn how much oil is naturally present in these food-materials; it will then be found that *ghi*, or some kind of expressed vegetable oil (such as sesamum, til, or mustard) must, in the majority of cases, be added in order to supplement the deficiency of this constituent. This addition of oil involves the derangement of the nutrient-ratio (that is, it lowers the proportion of albuminoids to carbonaceous nutrients) in any mixture of pulse and cereals, and therefore necessitates the withdrawal of a part of the cereal and the augmentation of the pulse. True, this change causes a reduction of the amount of oil already present in the ration, but this reduction is small, and may be neglected. In fact, we may be satisfied with the roughest approximations to the standard or theoretical figures in spite of the appearance of minute accuracy which three places of decimals give to our tables and to our calculated examples of rations. And it should be here mentioned that a margin of excess should be allowed in all weighings to compensate for inferiority of the grains, for casual impurities, and for the presence in an unavailable form of some of the nutrients in the food-staples employed. It is difficult to give any general rule, or to say what amount of allowance should usually be made; 5 per cent. will probably suffice, except in the case of obviously inferior or dirty samples.

There are many ways in which such tables as the two which we now give may be used in order to ascertain what amounts, in ounces and decimals of an ounce, of two or more foods, must be associated in order to furnish the necessary quantities of the two chief nutrients for a day's ration. We believe that the simple "inspection" method adopted in the examples which follow will be generally preferred to elaborate formulæ and equations involving much calculation, many references to tables, and the use of a number of coefficients.

TABLE I.

ALBUMINOIDS AND STARCH-EQUIVALENTS IN CEREALS AND BUCKWHEATS.

[*The figures represent ounces and decimals of an ounce.*]

		1 oz.	2 oz.	3 oz.	4 oz.	5 oz.	6 oz.	7 oz.	8 oz.	9 oz.
Ragí ...	Albuminoids	·059	·118	·177	·236	·295	·354	·413	·472	·531
p. 89	Starch ...	·764	1·528	2·292	3·056	3·820	4·584	5·348	6·112	6·876
Koda ...	Albuminoids	·07	·14	·21	·28	·35	·42	·49	·56	·63
p. 40	Starch ...	·82	1·64	2·46	3·28	4·10	4·92	5·74	6·56	7·38
Rice ...	Albuminoids	·073	·146	·219	·292	·365	·438	·511	·584	·657
p. 73	Starch ...	·797	1·594	2·391	3·188	3·985	4·782	5·579	6·376	7·173
Sánwa ...	Albuminoids	·084	·168	·252	·336	·420	·504	·588	·672	·756
p. 49	Starch ...	·794	1·588	2·382	3·176	3·970	4·764	5·558	6·352	7·146
Gundli ...	Albuminoids	·091	·182	·273	·364	·455	·546	·637	·728	·819
p. 44	Starch ...	·773	1·546	2·319	3·092	3·865	4·638	5·411	6·184	6·957
Maize ...	Albuminoids	·095	·190	·285	·380	·475	·570	·665	·760	·855
p. 65	Starch ...	·790	1·580	2·370	3·160	3·950	4·740	5·430	6·320	7·110
Joár ...	Albuminoids	·093	·186	·279	·372	·465	·558	·651	·744	·837
p. 80	Starch ...	·769	1·538	2·307	3·076	3·845	4·614	5·383	6·152	6·921
Shama ...	Albuminoids	·096	·192	·288	·384	·480	·576	·672	·768	·864
p. 50	Starch ...	·757	1·514	2·271	3·028	3·785	4·542	5·299	6·056	6·813
Bájra ...	Albuminoids	·104	·208	·312	·416	·520	·624	·728	·832	·936
p. 59	Starch ...	·791	1·582	2·373	3·164	3·955	4·746	5·537	6·328	7·119
Kángni ...	Albuminoids	·108	·216	·324	·432	·540	·648	·756	·864	·972
p. 55	Starch ...	·801	1·602	2·403	3·204	4·005	4·806	5·607	6·408	7·209
Barley ...	Albuminoids	·115	·230	·345	·460	·575	·690	·805	·920	1·035
p. 99	Starch ...	·730	·460	2·190	2·920	3·650	4·380	5·110	5·840	6·570
Chena ...	Albuminoids	·126	·252	·378	·504	·630	·756	·882	1·008	1·134
p. 43	Starch ...	·777	1·554	2·221	3·108	3·885	4·662	5·439	6·216	6·993
Amaranth...	Albuminoids	·143	·286	·429	·572	·715	·858	1·001	1·144	1·287
p. 108	Starch ...	·760	1·520	2·280	3·040	3·800	4·560	5·320	6·080	6·840
Wheat ...	Albuminoids	·135	·270	·405	·540	·675	·810	·945	1·080	1·215
p. 95	Starch ...	·712	1·424	2·136	2·848	3·560	4·272	4·984	5·696	6·408
Buckwheat..	Albuminoids	·152	·304	·456	·608	·760	·912	1·064	1·216	1·368
p. 114	Starch ...	·714	1·428	2·142	2·856	3·570	4·284	4·998	5·712	6·426
Quinoa ...	Albuminoids	·192	·384	·576	·768	·960	1·152	1·344	1·536	1·728
p. 110	Starch ...	·578	1·156	1·734	2·312	2·890	3·468	4·046	4·624	5·202

TABLE II.

ALBUMINOIDS AND STARCH EQUIVALENTS IN PULSE.

[*The figures represent ounces and decimals of an ounce.*]

		1 oz.	2 oz.	3 oz.	4 oz.	5 oz.	6 oz.	7 oz.
Soy-beans ...	Albuminoids	·353	·706	1·059	1·412	1·765	2·118	2·471
p. 143	Starch ...	·694	1·388	2·082	2·776	3·470	4·164	4·858
Vetchlings ...	Albuminoids	·319	·638	·957	1·276	1·595	1·914	2·233
p. 132	Starch ...	·521	1·042	1·563	2·084	2·605	3·126	3·647
Lupines ...	Albuminoids	·317	·634	·951	1·268	1·585	1·902	2·219
p. 123	Starch ...	·452	·904	1·356	1·808	2·260	2·712	3·164
Vetches ...	Albuminoids	·315	·630	·945	1·260	1·575	1·890	2·205
p. 131	Starch ..	·497	·994	1·491	1·988	2·485	2·982	3·479
Guár	Albuminoids	·298	·596	·894	1·192	1·490	1·788	2·086
p. 124	Starch ...	·494	·988	1·482	1·976	2·470	2·964	3·458
Lentils... ...	Albuminoids	·249	·498	·747	·996	1·245	1·494	1·743
p. 139	Starch ..	·595	1·190	1·785	2·380	2·975	3·570	4·165
Pea-nuts ...	Albuminoids	·245	·490	·735	·980	1·225	1·470	1·715
p. 127	Starch ...	1·267	2·534	3·801	5·068	6·335	7·602	8·869
Moth	Albuminoids	·238	·476	·714	·952	1·190	1·428	1·666
p. 152	Starch ...	·580	1·160	1·740	2·320	2·900	3·480	4·060
Peas	Albuminoids	·236	·472	·708	·944	1·180	1·416	1·652
p. 135	Starch ...	·575	1·150	1·725	2·300	2·875	3·450	4·025
Catiang-beans ...	Albuminoids	·231	·462	·693	·924	1·155	1·386	1·617
p. 156	Starch ...	·578	1·156	1·734	2·312	2·890	3·468	4·046
Haricots ...	Albuminoids	·230	·460	·690	·920	1·150	1·380	1·610
p. 147	Starch ...	·576	1·152	1·728	2·304	2·880	3·456	4·032
Múng-beans ...	Albuminoids	·227	·454	·681	·908	1·135	1·362	1·587
p. 151	Starch ...	·608	1·216	1·824	2·432	3·040	3·648	4·276
Horse-gram ...	Albuminoids	·225	·450	·675	·900	1·125	1·350	1·575
p. 162	Starch ...	·603	1·206	1·809	2·412	3·015	3·618	4·221
Lablab-beans ...	Albuminoids	·224	·448	·672	·896	1·120	1·344	1·568
p. 161	Starch ...	·574	1·148	1·722	2·296	2·870	3·444	4·018
Pigeon-peas ...	Albuminoids	·203	·406	·609	·812	1·015	1·218	1·421
p. 169	Starch ...	·596	1·192	1·788	2·384	2·980	3·376	4·172
Chick-peas ...	Albuminoids	·195	·390	·585	·780	·975	1·170	1·365
p. 128	Starch ...	·643	1·286	1·929	2·572	3·215	3·858	4·501
Inga beans ...	Albuminoids	·176	·352	·528	·704	·880	1·056	1·232
p. 173	Starch ...	·807	1·614	2·421	3·228	4·035	4·842	5·649

Examples of Daily Rations.—Now suppose that we have to arrange a dietary of which soy-beans and rice are the substantive constituents. Keeping the quantity of the more costly and less digestible pulse as low as possible and neglecting for the moment the question of fat or oil, we shall find from the 4th and 5th columns of table 2 and from the 8th column of table 1, that 4½ ounces of soy-beans and 8 ounces of rice will supply the necessary amounts of albuminoids and starch :

	Albuminoids.	Starch.
Soy-beans, 4½ oz. furnish ...	1·589	3·123
Rice, 8 oz. furnish	·584	6·376
	2·173	9·499
Amounts required	2·128	9·250

The only remaining question relates to the fat or oil of this ration. It so happens that soy-beans are rich in oil, the above quantity, of 4½ ounces, containing ·851 of an ounce of oil. Now the theoretical quantity demanded by this ration is ·752 of an ounce—so that in this case no additional oil or fat is wanted. We will take another example, employing a ration in which rice is associated with a kind of pulse in which there is little more than a trace of oil, and where the albuminoids are present in no more than the ordinary proportion ; such a pulse is the lablab-bean. Adopting the same plan of extracting figures from tables 1 and 2 as before, we find that—

	Albuminoids.	Starch.
Lablab-beans, 8 oz. furnish ...	1·792	4·712
Rice, 6½ oz. furnish	·438	4·782
	2·230	9·494
Amounts required	2·128	9·25

But there is no appreciable amount of oil in this ration ; it must be supplied either by an oily seed, or by a separated oil or fat. Assuming the latter plan to be adopted, we add ¾ ounce oil, reducing the rice by 2½ ounces and increasing the lablab-beans by ½ ounce ; the latter step is rendered necessary by the obligation of replacing in a more concentrated form the albuminoids withdrawn in the 2½ ounces of rice. The ration will then stand—

	Albuminoids.		Starch.
Lablab-beans, 8½ oz. furnish ...	1·904	...	4·999
Rice, 4 oz. furnish	·292	...	3·188
¾ oz. Oil furnishes	*nil*	...	1·725
	2·196	...	9·912

These totals approach sufficiently near to the figures required by theory.

We will not multiply examples of dietaries containing rice associated with pulse, but will consider the cases furnished by millets and pulse used together. Happily, the comparative richness of the millets in albuminoids enables us to lower the amount of lablab or ordinary pulse which it is necessary to employ with them in constructing a standard ration. As an illustration we will use múng-beans and bájra (*Pennisetum typhoideum*). By reference to the tables it will be found that something like these quantities are required :

	Albuminoids.		Starch.
Múng-beans, 6 oz. furnish ...	1·362	...	3·648
Bájra, 8 oz. furnish	·832	...	6·328
	2·194	...	9·976

These form practically the ration required. But there is an insufficient amount of oil in this dietary ; adding ¾ ounce of oil, we shall have to withdraw 2 ounces of bájra and to increase the múng by ½ ounce. The figures become :

	Albuminoids.		Starch.
Múng-beans, 6½ oz. furnish ...	1·587	...	4·276
Bájra, 6 oz. furnish	·520	...	3·955
¾ oz. Oil equal to	*nil*	...	1·725
	2·107	...	9·956

So far we have been considering dietaries belonging to the lowest or bare-sustenance scale ; the scales for moderate work and heavy work have now to be similarly discussed.

We begin with the dietary for moderate work, still assuming that the body-weight of the persons to be fed is 105 lbs., and employing the same pulses and the same cereals as in the three examples of rations previously constructed. Referring, as before, to the tables, we find that—

	Albuminoids.	Starch.
Soy-beans, 5 oz. furnish ...	1·765 ...	3·470
Rice, 16 oz. furnish	1·163 ...	12·752
	2·928 ...	16·222
Amounts required... ...	2·954 ...	15·779

As the oil contained in this ration is but 1 ounce and should be 1·4 ounces, it will be necessary to add the lacking 4-10ths. To do this without disturbing the ratio, 1½ ounces of rice should be withdrawn, and ½ ounce of soy-beans added. Thus the amounts will be :

	Albuminoids.	Starch.
Soy-beans, 5½ oz. furnish ...	1·942 ...	3·817
Rice, 14½ oz. furnish	·954 ...	11·552
Oil, $\frac{4}{10}$ oz. equal	*nil* ...	·920
	2·896 ..	16·289

If rice and lablab-beans be employed, the following amounts will be necessary :

	Albuminoids.	Starch.
Lablab-beans, 8 oz. furnish ...	1·792 ...	4·592
Rice, 16 oz. furnish	1·168 ...	12·752
	2·960 ...	17·344

As, however, this dietary should include 1·41 ounces of oil, it will be necessary to supplement its natural deficiency in this constituent by adding at least 1 ounce. We withdraw, therefore, 4 ounces of rice, and add 1 ounce of lablab-beans.

	Albuminoids.	Starch.
Lablab-beans, 9 oz. furnish ...	2·016 ...	5·166
Rice, 12 oz. furnish	·876 ...	9·664
Oil, 1 oz. equal to	*nil* ...	2·300
	2·892 ...	17·130

In truth, a still larger proportion of lablab would be needed to realise the proper ratio between the constituents of this dietary; but 9 ounces of pulse is already too large a quantity to be consumed daily, and we ought, wherever possible, to associate a third aliment with the two employed. For example, wheat or one of the millets might advantageously replace a part of the rice.

In the third example of a moderate-work ration, múng-beans and bájra are assumed to be employed.

	Albuminoids.		Starch.
Múng-beans, 5½ oz. furnish ...	1·249	...	3·344
Bájra, 16 oz. furnish	1·664	...	12·656
	2·913	...	16·000

These quantities of múng and bájra contain about ·65 ounce of oil; this leaves ·75 to be otherwise supplied. If ¾ ounce be introduced, 3 ounces of bájra must be withdrawn and 1½ ounces múng added:

	Albuminoids.		Starch.
Múng-beans, 7 oz. furnish ...	1·587	...	4·276
Bájra, 13 oz. furnish	1·352	...	10·183
Oil, ¾ oz. equal to	*nil*	...	1·725
	2·939	...	16·184

In the third dietary scale, that of hard work, a single example will probably suffice to show the mode of constructing the normal ration; soy-beans and bájra shall be taken, and the small necessary addition of oil made.

	Albuminoids.		Starch.
Soy-beans, 6½ oz. furnish ...	2·295	...	4·511
Bájra, 13½ oz. furnish ...	1·404	...	10·578
Oil, ¾ oz. equals	*nil*	...	1·725
	3·699	...	16·814
Amounts required	3·635	...	16·954

The only dietaries, other than Indian, which are made up of similar foods and have been chemically examined, are those of Japan prisons. In these, however, no attempt seems to have been made to adjust the composition of the rations to the work required, for the amounts of the pulse, seaweed, and other vegetables rich in nitrogen, are the same for all the classes, the quantities of rice or rice and barley given being augmented. As the body-weights of the Japanese prisoners do not differ much from those in Indian jails, we may usefully quote here the rations of such a prison as that of Tokio.

Ration.	Albuminoids.	Fat.	Starch.	Nutrient-ratio.
a. No work ...	... 1·7 ...	·24 ...	12·8 ...	1 : 7·9
b. Light labour	... 2·0 ...	·27 ...	16·1 ...	1 : 8·1
c. Hard labour	... 2·9 ...	·32 ...	21·7 ...	1 : 8·6

The small quantities of albuminoids and of fat in these rations, and the excessive amount of starch, are very noticeable. In dietary *c* the nutrient-ratio of 1 : 8·6 should be contrasted with that of C given on p. 20, namely, 1 : 4·66.

The following are additional standard dietaries, reduced in the case of rations for adults, to a mean body-weight of 105 lbs.

	Albuminoids.	Fat.	Starch, etc.	Nutrient-ratio.
I. Children of 6 to 15 years...	2·7 ...	1·3 ...	8·7 ...	1 : 4·33
II. Ordinary labour	3·12 ...	2·34 ...	10·37 ...	1 : 5·04
III. " "	3·15 ...	1·2 ...	14·02 ...	1 : 5·33

The amount of mineral matter, including common salt, required per day, is just under 1 ounce.

It should be observed that the amount of oil or fat in the dietary III. just given, is rather lower than is desirable. It could not, however, be increased unless a considerable diminution of the starch, already too high, were also effected; of course, the amount of albuminoids would then have to be raised as the proportion in which they exist in this dietary is low.

Many other standard dietaries have been proposed from time to time, in addition to those already given in previous paragraphs. On comparing them with all or nearly all the public dietaries in actual and extensive use, the latter show a deficiency of albuminoids and fat, and an excess of carbohydrates. The great cost of butchers' meat and the difficulty of obtaining regular supplies of fish are the chief reasons why these defects have not been generally remedied. In many countries some improvement in this direction has been effected by the more abundant employment of pulse in the daily rations. As an illustration of the divergencies between an actual and a standard dietary we may cite the case of the average European soldier:

	Actual.	Standard.
Albuminoids	4·0 oz.	5·0 oz.
Fat	1·25 ,,	3·0 ,,
Carbohydrates	18·0 ,,	14·0 ,,
Nutrient-ratio =	1 : 5·25	1 : 4·25
Nutrient-value =	24·9 oz.	25·9 oz.

PART III.

THE CEREALS.

THE grain of the grasses is a complete fruit; in a few familiar cases (barley) it commonly retains as an additional and firmly-attached covering a pair of paleæ. In all cases where these floral envelopes adhere to the winnowed grain they must be removed by some mechanical operation before the cereal can be used as human food; the harshness of these envelopes as well as the indigestibility of the large amount of fibre which they contain necessitate this treatment.

A considerable and often excessive proportion of starch is a characteristic feature of the composition of these grains. This constituent is present in largest proportion in the two millets, ragí and koda, and in cleaned, that is, husked rice; in wheat, and many of the millets, the percentage of this nutrient is lower. Most of the larger cereal grains, which have been tested for sugar, have been found to contain a small proportion of that compound, or rather, at least two different kinds of sugar have been found in them, although the total amount of saccharine substance in well-ripened corn probably does not exceed 2 per cent. For the purposes of this work it is quite unnecessary to distinguish between such sugar and the starch present in these grains; in consequence, the starch and sugar with the dextrin, gum, and similar matters will always be found, in the following pages, under the single heading of "starch." It must not, however, be supposed that the starch of all these grains is absolutely identical in feeding value, for this substance, as extracted from different grains, is found to vary in the ease with which it is affected by hot water and by those digestive ferments which are capable of dissolving it during its passage through the alimentary tract.

The proportion of flesh-forming substances, or albuminoids, in different cereals, varies much — ranging from 18 per cent. in the hardest wheats to 6·5 per cent. in the poorest rice. In different varieties of many, perhaps most cereals, this constituent does not oscillate between very wide limits; but in wheat the range is very great. In Indian wheat, however, the limit of variation in the percentage of albuminoid matter does not seem to exceed 6½ per cent., that is from 10·3 to 16·7 per cent. There are considerable differences in the albuminoids present in different cereals, the albuminoids of wheat presenting more analogy to animal fibrin, and being more easy to separate and more tenacious, stringy, and elastic when separated from the other constituents of the grain than is the case with other cereals. Upon these peculiarities of wheat-albuminoids depends the power of forming a light vesiculated bread which wheat-flour possesses in an eminent degree. It is usual to calculate the amount of albuminoids in the cereal grains by multiplying the amount of nitrogen which analysis shows to exist in them by the coefficient 6·3. The results thus obtained are always too high, for the cereal grains invariably contain nitrogen in the form of compounds not albuminoid and not possessing the same functions in nourishing the body. But a large number of careful estimations of the actual albuminoids in well-ripened grains of the most familiar cereals has been made, with the result of showing that the amount to be deducted from the total calculated percentages of albuminoids may generally be neglected without serious error. Were such deductions allowed they would amount to from 1½ to 9 parts from each 100 parts of albuminoids. As in most Indian-grown grains and seeds no such determinations of non-albuminoid nitrogen have been yet made, it would have been impossible to have corrected the figures in the following pages in accordance with the more exact method. As the author of the present Handbook was the first chemist who devised an analytical method for distinguishing between albuminoid and non-albuminoid nitrogen in foods, he may be permitted to add that exaggerated deductions have been made in some published works on food and dietetics for non-albuminoid nitrogen. And he would further urge that, as many theoretical or standard dietaries have been constructed in part upon the old and incorrect data as to the albuminoids in the foods consumed, such dietaries may be not unfairly imitated by the use of analytical results obtained in the same, that is, the usual way. It would occupy too much space were all the evidence in support of this argument to be marshalled before our readers; but, taken in connection with the other reasons we have urged, we think it justifies the course we intend to follow in the present elementary Handbook.

There is good reason to conclude that the albuminoids of the cereals, while presenting, if pure, very slight variations in the proportions of their ultimate chemical elements, exist in a number of states or modifications. The so-called "gluten" of wheat may be resolved into three or possibly four constituents; two of these occur in maize, where they are accompanied by a third albuminoid not found in wheat and called "zein." These are not products obtained by the chemical treatment to which the wheaten flour and maize flour are submitted in the processes of extraction—they are *educts* pre-existing in the raw materials operated upon. The properties of these, and other members of the group of cereal albuminoids, are probably identical in one respect only, namely, their value as nutrients; their distinctive physical and mechanical properties are certainly in many cases well marked. Thus (as is stated on page 32) it is upon the peculiar elastic and viscous character of the chief albuminoid of wheat-grain, the gluten-fibrin, that its admirable appropriateness for the making of a light vesiculated bread depends. It may be here observed that those valuable mineral salts, the alkaline and earthy phosphates, are very intimately associated with the albuminoids, although the union can hardly be regarded as a chemical one. And it has often been observed that the larger the percentage of albuminoid matter in a grain the larger is the proportion of these phosphates.

It is necessary to add a few words as to the oil or fat present in the cereals. Some of this, often a good proportion, resides in the coats of the grains, and in the embryos or "chits." When, therefore, wheat-grain is decorticated, and especially when, as in some modern processes of milling, the embryos are removed, the percentage of oil is much lowered. This will be obvious when we state that the embryos contain 14¼ per cent. of oil and the different grades of bran from 2¼ to 5, the whole wheat-grain not showing more than 2 per cent. Maize and barley are rich in oil—rice very poor. In actual dietaries this deficiency of oil or fat is made up in several ways. Where fish or flesh meat is employed, a good deal of oil or fat is introduced therein, but the very common use in India of milk, or of some preparation from it, effects something in the same direction; expressed fatty oils are also extensively employed.

The mineral matter or ash in the grain of the cereals shows some variation both in quantity and in composition. In the unhusked grains it amounts on the average to less than 2 per cent.; in cleaned rice it is as low as 0·6 per cent., and may occasionally sink to 0·4. In paddy it amounts to 0·7 or 0·8 per cent.; as a general rule, the removal of the coats of any cereal grain in dressing it, involves the abstraction of much mineral matter.

The large group of the minor cereals, which may be designated "millets," together constitute a more important crop than either rice or wheat, and are grown more extensively, being raised from Madras in the South to Rájputána in the North. They occupy about 83 per cent. of the food-grain area in Bombay and Sind; 41 per cent. in the Punjab; 39 per cent. in the Central Provinces; and 34 per cent. in the North-West Provinces. The chief species are Great Millet, or joár (*Sorghum vulgare*); spiked or bulrush millet, bájra (*Pennisetum typhoideum*); Ragí (*Eleusine coracana*);

Italian millet (*Setaria italica*); and Chena (*Panicum miliaceum*). Besides these, other kinds are cultivated, such as Panicum miliare, and Paspalum scrobiculatum, while the grain of a few wild kinds is used in times of scarcity.

The following table shows the number of acres in 1882–83 under two of the chief species of millet (Joár and Bájra) in five Provinces from which the returns are fairly complete:

	Joár.	Bájra.
Madras	3,691,127 ...	2,496,958
Bombay	5,392,469 ...	4,566,113
Punjab	2,552,248 ...	3,218,248
N.W. Provinces	2,828,375 ...	2,009,298
Berar	2,276,220 ...	92,322

More recently, the total area under millets has been estimated at 33¼ million acres.

As a rule the millets are all *Kharif*, or autumnal harvest crops, being generally sown in the early weeks of the monsoon, and in June or July, and reaped in October and November.

Next in importance to the millets comes the group of the larger cereal grains. This includes wheat, rice, barley, and maize. The importance of the Indian wheats and of rice is discussed with some degree of fulness in the pages of the Handbook devoted to these cereals. While wheat approaches in composition that of a standard food, rice is very far from doing so, both in regard to albuminoids and mineral matters. Much more pulse for example, must be introduced into a rice diet than into one of wheat or into one of millet, in order that the chemical proportions of the several nutrients required for a perfect food may be reached.

The chief methods of preparing or cooking the cereal grains for the food of man are mentioned under the heads of the several kinds to which they more particularly apply. (See, for example, under Rice, p. 73, Maize, p. 66, Bulrush-millet, p. 59.)

The following table shows the number of acres, in 1882–83, under the two chief larger cereals, namely wheat and rice, in five

Provinces where both crops are important and whence fairly complete returns have been obtained :

	Wheat.	Rice.
Punjab	6,734.357 ...	775,367
N.W. Provinces	4,676,580 ...	3,055,933
Central Provinces ...	3,619,704 ...	4,416,054
Oudh	1,863,753 ...	1,743,015
Bombay	1,630,310 ...	1,874,934

As a general rule wheat and barley are *Rabi*, or winter harvest crops, being sown at the end of the monsoon and reaped between January and March.

The Order Gramineæ or Grasses contains between 3,100 and 3,200 species. A conspectus of the tribes and genera of this Order—so far as the plants described in the following pages are concerned—is here given :

Order, GRAMINEÆ.

Tribes.	Genera.
i. PANICEÆ.	Paspalum, Panicum, Oplismenus, Setaria, Cenchrus, Pennisetum.
ii. MAYDEÆ.	Coix, Zea.
iii. ORYZEÆ.	Oryza.
vi. ANDROPOGONEÆ.	Saccharum, Sorghum.
ix. AVENEÆ.	Avena.
x. CHLORIDEÆ.	Eleusine.
xii. HORDEÆ.	Triticum, Hordeum.
xiii. BAMBUSEÆ.	Bambusa.

The sequence in which the several kinds of Indian cereals and grain-substitutes are arranged and described in the following pages is that indicated in the above table. A few species have been, however, relegated to the closing paragraphs of the present Part, because our chemical information concerning them is either very meagre or absolutely wanting.

In classifying the several kinds of cereals according to their composition the most useful plan to follow is one based on their relative richness in albuminoids. This richness is best expressed in the form previously explained (p. 14) under the name "nutrient-ratio," or the proportion between the albuminoid they contain and their starch, reckoning, however, with their starch, the starch-

38 FIG. 1. KODA MILLET (*Paspalum scrobiculatum*).

equivalent of the oil present. In the following table the first column of figures give the nutrient-ratio, the second the percentage of albuminoids, and the third, the nutrient-value, or the added percentage of albuminoids, of starch, and of oil translated into its starch-equivalent.

Name of Cereal.	Nutrient-ratio, or Albuminoids to Starch.	Albuminoids, percentage of.	Nutrient-value.
1. Eleusine coracana...	1 : 13	5·9	84
2. Paspalum scrobiculatum	1 : 11·7	7·0	89
3. Oryza sativa	1 : 10·8	7·3	86½
4. Panicum frumentaceum	1 : 9·5	8·4	88
5. Panicum miliare	1 : 8·4	9·1	85
6. Zea Maÿs	1 : 8·3	9·5	88½
7. Sorghum vulgare	1 : 8·2	9·3	86
8. Panicum colonum...	1 : 8	9·6	85
9. Pennisetum typhoideum	1 : 7·6	10·4	89½
10. Setaria italica	1 : 7·4	10·8	91
11. Sorghum saccharatum	1 : 6·4	11·8	87
12. Bambusa arundinacea	1 : 6·4	11·8	87
13. Hordeum vulgare...	1 : 6·3	11·5	84½
14. Panicum miliaceum	1 : 6	12·6	89
15. Triticum sativum ...	1 : 5·2	13·5	84½

It will be seen from the above tabular statement that Indian wheat (15), and Indian millet (14), present a nutrient-ratio not far removed from that of a standard diet (1 : 5); while Ragí (1), and Koda Millet (2), and Rice (3) are very deficient in their proportion of albuminoids or flesh-formers.

KODA MILLET.

Paspalum scrobiculatum, L.

Sanskrit—Korádusha, Kodrava.
Hind.—Kodaka. *Beng.*—Koda. *Punjab*—Kodra. *North-West Provinces*—Kodon, Marsi. *Telugu*—Allu. *Gujrat*—Menya.
Bombay—Kodra, Harik. *Sinhalese*—Wal-amu. *Tamil*—Waragu.

This erect grass, which grows to a height of 1 to 2 feet, is frequent in tropical and sub-tropical Asia, and is largely grown, especially on poor soils. The seed is sown in June and July, the harvest takes place in October and November. It is used in the districts of Mirzapur, Gorakhpur, Patná, Sháhábád,

Sáran, Bhágulpur, Purniah, and Santál Parganás. It is largely grown on the lowlands near the Ganges in Bhágulpur. The analysis of the grain without husk is here given:

COMPOSITION OF KODA MILLET (HUSKED).

	In 100 parts.	In 1 lb.
Water	11·7 ...	1 oz. 382 grs.
Albuminoids	7·0 ...	1 ,, 52 ,,
Starch	77·2 ...	12 ,, 154 ,,
Oil...	2·1 ...	0 ,, 147 ,,
Fibre	0·7 ...	0 ,, 49 ,,
Ash	1·3 ...	0 ,, 91 ,,

The nutrient-ratio is here 1 : 11·7, and the nutrient-value 89.

It is said, apparently with truth, to be at times the cause of vertigo, and is not considered to be as digestible as Setaria italica; the stems afford an inferior fodder. Koda is boiled and eaten in the same way as rice, or else is parched and ground, the meal being made into a kind of pudding. The alleged comparative indigestibility of this grain cannot be attributed to its percentage of fibre, which is unusually low, but must be owing to some constituent or some condition of the grain which ordinary chemical analysis does not reveal.

Mr. Duthie says of this plant that it is the most largely-grown (in the North-West Provinces and Oudh) of all the lesser millets, being a favourite crop for inferior out-lying land. He adds that it is not a popular article of food; the yield may amount to 10 or 12 maunds per acre, but much of this is chaff.

CHENA OR INDIAN MILLET.

Panicum miliaceum, L.

Synonyms—P. asperrimum (Lagasc.); P. Milium (Pers.).

Sanskrit—Vrihibheda. *Hind.*—Chena, Chína. *Beng.*—Chennah. Other vernacular names are: Sáwan-jethwa, Kuri, Phikar, Ráli, Bausi. In *Tamil*—Varagu, Katacuny.

This millet is an erect annual about 2 feet high, with rather broad hairy leaves and a much divided nodding panicle. It was early introduced and is largely grown in many parts of India;

 FIG. 2. INDIAN MILLET (*Panicum miliaceum*).

it has become naturalised in Ceylon. It generally needs but little rain. It ascends the Himálaya as far as 10,000 feet. It is extensively cultivated in some parts of Purniah District, and is used also in Bhágulpur, Patná, and Champáran, the yield being considerable. It is a quick-growing plant,* a second crop being often obtained from the grain shed in harvesting the first. It proves a useful crop in times of scarcity, not only from its rapid growth, but from its flourishing in seasons of comparatively small rainfall. Mr. Duthie, however, states that in the districts on which he reports, this crop needs a good deal of watering, and that the yield does not amount to more than from 6 to 8 maunds per acre. In the Deccan it is sown in June and July, and reaped in November and December.

COMPOSITION OF CHENA MILLET (HUSKED).

	In 100 parts.	In 1 lb.
Water	12·0	1 oz. 402 grs.
Albuminoids	12·6	2 ,, 7 ,,
Starch	69·4	11 ,, 45 ,,
Oil	3·6	0 ,, 252 ,,
Fibre	1·0	0 ,, 70 ,,
Ash	1·4	0 ,, 98 ,,

The nutrient-ratio is here 1 : 6, and the nutrient-value 89.

It should be noted here that a European sample of this millet, analysed without the husk, gave 15·3 per cent. of albuminoids, 61·6 per cent. of starch, 5 per cent. of oil, and 3·5 per cent. of fibre; these figures correspond to a nutrient-ratio of 1 : 4·8, and a nutrient-value of 88, and would lead to a still higher estimate of the worth of this food.

This millet is boiled and eaten with milk and sugar, or with dal in curries, or as "mara," the grains being slightly boiled, dried, parched in hot sand, and then sifted from the husks; the "mara" is then eaten with sour milk—this preparation is a favourite food at marriage-feasts.

* Mr. J. A. Baines states, however, that its growth is slower than that of Italian millet.

LITTLE MILLET.

Panicum miliare, Lam.

Synonyms—P. attenuatum (Nees.); Milium effusum (Lour.).
Hind.—Kutki, Gundli. *Tamil*—Chámai. *Sinhalese*—Menéri. *Beng.*—Gondula. *Punjab*—Kutki. *Telugu*—Nella-shama. *Bombay*—Warai. *Madras*—Shama.

An annual grass, 2 to 3 feet high, producing many erect branching culms from a single seed. The panicle becomes beautifully bowed with the increasing weight of the grain. This species is cultivated to some extent in many districts; it is grown in the Punjab up to the Kheri Pass; also in Nepal and Central India.

COMPOSITION OF LITTLE MILLET.

	In 100 parts.	In 1 lb.
Water	10·2 ...	1 oz. 277 grs.
Albuminoids	9·1 ...	1 „ 199 „
Starch	69·0 ...	13 „ 17 „
Oil...	3·6 ...	0 „ 252 „
Fibre	4·6 ...	0 „ 322 „
Ash	3·5 ...	0 „ 245 „

The nutrient-ratio is here 1 : 8·4, and the nutrient-value 85. Cattle are fond of the straw of this grass.

FIG. 3. LITTLE MILLET (*Panicum miliare*).

FIG. 4. SÁNWA MILLET (*Panicum frumentaceum*).

SÁNWA MILLET.

Panicum frumentaceum, Roxb.

Synonyms—Oplismenus frumentaceus (Kunth.); Echinochloa frumentacea (Link.). *Hind.*—Sánwa, Sámuka, Sawa. *Beng.*—Sama, Kheri. *Deccan*—Kathlí. *Sind.*—Sarou. *Kashmir*—Karin. *Telugu*—Bouth-shama. *Sinhalese*—Wel-marukku.

The culms of this millet are erect, from 2 to 4 feet high; the leaves are large and the panicle erect, oblong, and rigid. It is composed of numerous incurved spikes.

Sánwa is the quickest growing of all the millets; the harvest may take place within 6 weeks of the sowing. Where it is sown in April and May it is cut in June and July; the June sowing is ready in August. As it ripens thus quickly and early, it affords a supply before the bájra harvest begins. Mr. Duthie, speaking of the North-West Provinces and Oudh, estimates the yield of grain as 8 to 10 maunds per acre, but on very poor soil it may be as low as 4.

COMPOSITION OF SÁNWA (WITH HUSK).

	In 100 parts.	In 1 lb.
Water	12·0 ...	1 oz. 402 grs.
Albuminoids	8·4 ...	1 „ 150 „
Starch	72·5 ...	11 „ 263 „
Oil...	3·0 ...	0 „ 210 „
Fibre	2·2 ...	0 „ 154 „
Ash	1·9 ...	0 „ 133 „

The nutrient-ratio is here 1 : 9·5, and the nutrient-value 88.

Sánwa does not take a high place among the millets. It is either boiled as rice, or boiled with milk, and eaten with sugar, or it is parched.

SHAMA MILLET.

Panicum colonum, L.

Synonym—Oplismenus colonus (Kunth.).
Beng.—Shama. *North-West Provinces*—Sarwak, Shamak.

This millet is common in the tropical and sub-tropical regions of the Old World—it ascends to 6,000 feet in the Himálaya. The panicle contains 8 or 10 one-sided spikes, each ½ to ¾ inch long. The stems are erect or decumbent at the base, about 2 feet or 3 feet high.

COMPOSITION OF SHAMA (HUSKED).

	In 100 parts.	In 1 lb.
Water	12·0 ...	1 oz. 403 grs.
Albuminoids	9·6 ...	1 ,, 234 ,,
Starch	74·3 ...	11 ,, 388 ,,
Oil...	·6 ...	0 ,, 42 ,,
Fibre	1·5 ...	0 ,, 105 ,,
Ash	2·0 ...	0 ,, 140 ,,

The nutrient ratio is here 1 : 8, and the nutrient-value 85.

This millet, sometimes called "Wild Rice," or "Jungle Rice," is a poor food; it is used, however, in some places where it grows freely (*e.g.*, in Champáran), by a considerable number of labourers as a usual article of diet. It is very good fodder-grass.

Other species of Panicum are occasionally, especially in times of drought, distress, or famine, consumed as food. Amongst these may be named P. flavidum (Retz.); P. fluitans (Retz.) (the common semi-aquatic species of Panicum); and P. hydaspicum (Edge.). The last species grows wild in the Punjab; Edgeworth mentions that its seeds are swept up from the ground and are eaten by the poor. Another species, P. psilopodium ("kutki," "mijhri") is mentioned by Mr. Duthie as grown on the poorest land and yielding about 2 maunds of seed per acre. Other species named by the same authority are—P. crus-galli (L.) ("bharti"), yielding an edible grain and good fodder, and P. Helopus (Trin.) ("kuri"), yielding an edible grain.

FIG. 5. SHAMA MILLET (*Panicum colonum*).

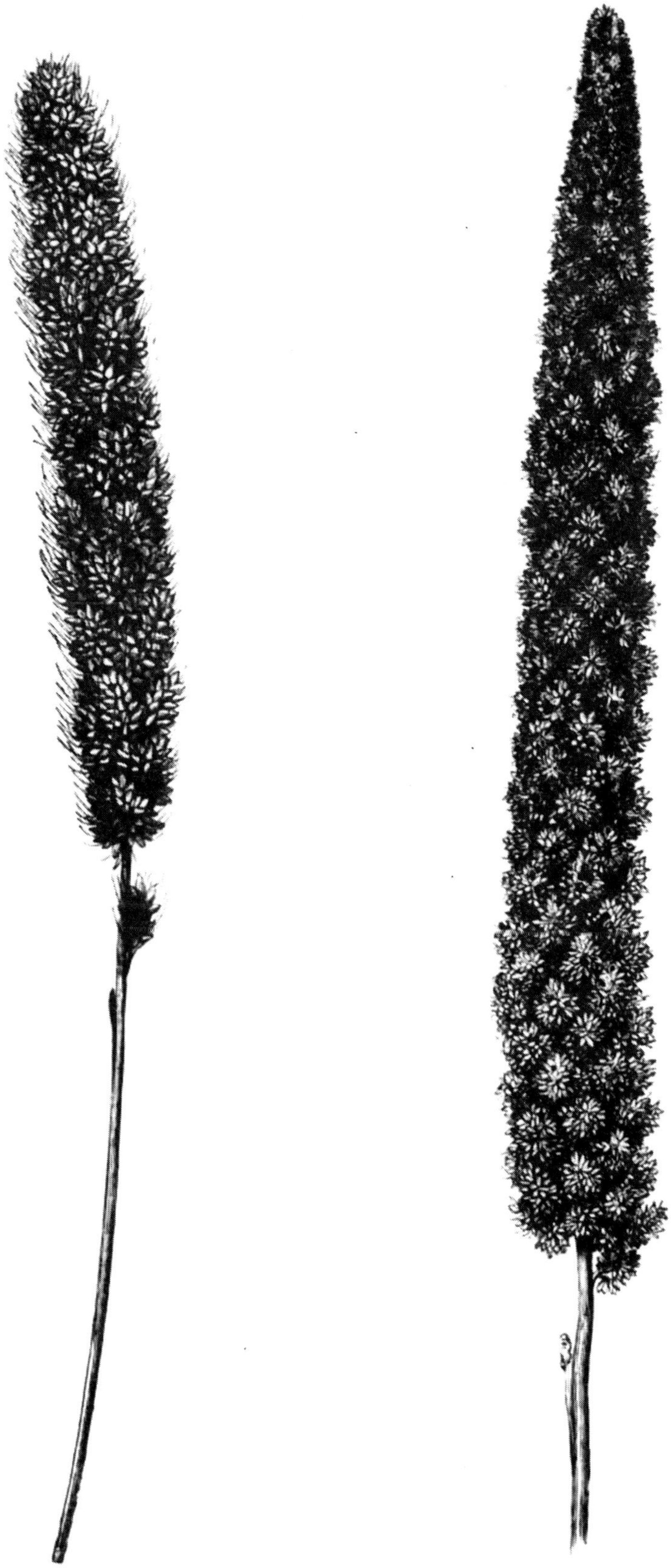

 FIGS. 6 AND 7. ITALIAN MILLET (*Setaria italica*). Two Varieties.

ITALIAN MILLET.

Setaria italica, Beauv.

Synonyms—Panicum italicum (L.); Pennisetum italicum (Brown).
Hind.—Kángni, Tángan, Kayuni, Kungú, Rawla. *Beng.*—Kákun. *Sinhalese*—Tana-hál. *Tamil*—Tinai. *Tel.*—Kora.
Sanskrit—Kángu, Priyangu.

This annual grass has a cylindrical panicle, at length nodding at the summit. It is frequently grown as an intermediate or subordinate crop; in some districts it is sown in April and May, and cut in June and July; in others it is sown from June to September, and reaped from September to January. It is largely used as human food in Patná, Sáran, Bhágulpur, Dinájpur, and Purniah; in some parts of Purniah it is extensively grown. Italian millet requires good land and is greatly benefited by manure.

COMPOSITION OF ITALIAN MILLET (HUSKED).

	In 100 parts.	In 1 lb.
Water	10·2	1 oz. 277 grs.
Albuminoids	10·8	1 „ 318 „
Starch	73·4	12 „ 63 „
Oil	2·9	0 „ 203 „
Fibre	1·5	0 „ 105 „
Ash	1·2	0 „ 84 „

The nutrient-ratio is here 1 : 7·4, the nutrient-value 91.

Italian millet is generally regarded as nutritious and digestible, but in some places it is considered to be rather heating. Sometimes it is boiled and eaten either alone or with the addition of milk and sugar (forming the preparation called "sír"). Sometimes it is parched. The percentage of flesh-forming matter in this grain seems to vary a good deal—from 9 to 13. The fibre in the unhusked grain may be as high as 8 per cent.

A species of Cenchrus (*C. echinatus*, L.), a genus allied to Setaria and Pennisetum, is not uncommon on the more arid parts of the Punjab plains. The grain is used for food in times of scarcity. Mr. Duthie remarks that Cenchrus catharticus (Del.) furnishes a grain used for food by the poor, who mix it with bájra.

BULRUSH MILLET.

Pennisetum typhoideum, Rich.

Synonyms—Penicillaria spicata (Willd.); Holcus spicatus (L.); Panicum spicatum (Roxb.).
Hind.—Bájra, Bájri, Lahra. *Tamil*—Cumbú. *Telugu*—Gantilú.

FIG. 8. BULRUSH MILLET (*Pennisetum typhoideum*).

An erect grass with thick round stems, generally 3 to 6 feet in height. The leaves are broad and long; the spikes are terminal, cylindric in form, and 6 to 9 inches long.

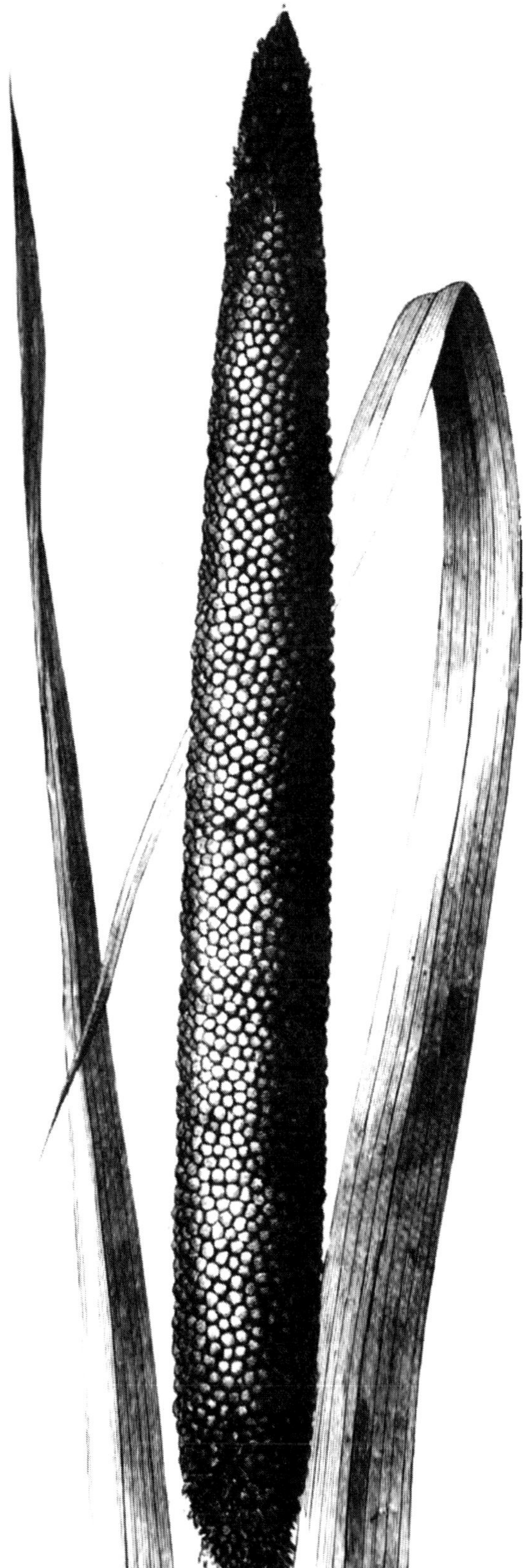

 FIG. 9. BULRUSH MILLET (*Pennisetum typhoideum*).

The seed is sown from June to August; the crop is cut from August to December, but chiefly between September and November. In the North-West Provinces and Oudh, where it is extensively grown (over 2 million acres), the plant often attains a height of 6 to 8 feet; the yield of grain is 5½ to 7 maunds.

COMPOSITION OF BULRUSH MILLET (UNHUSKED).

	In 100 parts.	In 1 lb.
Water	11·3 ...	1 oz. 354 grs.
Albuminoids	10·4 ...	1 „ 290 „
Starch	71·5 ...	11 „ 192 „
Oil...	3·3 ...	0 „ 231 „
Fibre	1·5 ...	0 „ 105 „
Ash	2·0 ...	0 „ 140 „

The nutrient-ratio is here 1 : 7·6, and the nutrient-value 89½.

This grain contains ·42 per cent. of potash and ·68 per cent. of phosphoric acid.

In Patná (where this millet is commonly eaten), it is considered somewhat heating; in Mirzápur it is used in the cold season by the poorer classes. It is ground into flour and made into cakes; or it is parched, coarsely ground, and then mixed with water, being eaten with gúr, curds, etc. The chopped stalks are good fodder.

Another Pennisetum, P. cenchroides (Rich.), is common in many parts of the Punjab where its seeds are swept up from the ground to be used as human food. Mr. Duthie names P. sanguinale (L.) as also furnishing an edible grain.

JOB'S TEARS.

Coix lachryma, L.

Synonyms—Coix arundinacea (Lamck.); Lithagrostis lachryma-Jobi (Gaertn.). *Hind.*—Kauch-gurgur, Saukrú, Lechusa. *Beng.*—Gurgur, Kunch. *Bombay*—Kassaibija. *Sinhalese*—Kíkir-rindí. *Naga Hills*—Re-sí.

An annual grass, occurring as a weed of cultivation on the rice-fields of Bengal. It is found on the plains of the Punjab, in the North Provinces, and on the warm hillsides and valleys of the Himálaya. It ascends to 5,000 feet on the Naga hills.

COMPOSITION OF JOB'S TEARS (HUSKED), C.

	In 100 parts.	In 1 lb.
Water	13·2 ...	2 oz. 49 grs.
Albuminoids	18·7 ...	2 „ 434 „
Starch	58·3 ...	9 „ 143 „
Oil...	5·2 ...	0 „ 364 „
Fibre	1·5 ...	0 „ 105 „
Ash	2·1 ...	0 „ 147 „

The nutrient-ratio is here 1 : 3·8, the nutrient-value 89.

Throughout Assam, and in the Eastern frontier-lands of India, this coarse cereal constitutes an important food of the hill tribes, replacing to some extent the millets of Northern and Southern India. In Burma the grains are eaten after having been parched like Indian corn. In the sample of this grain which gave the above analytical figures it was found that the edible seed after the removal of the hard and shining gray husk did not weigh more than 1 for every 4 parts by weight of the whole grain operated upon. The whole grains of the wild kinds are exceedingly hard and are used as beads; those of the cultivated variety are much softer and more easily husked.

 FIG. 10. JOB'S TEARS (*Coix lachryma*).

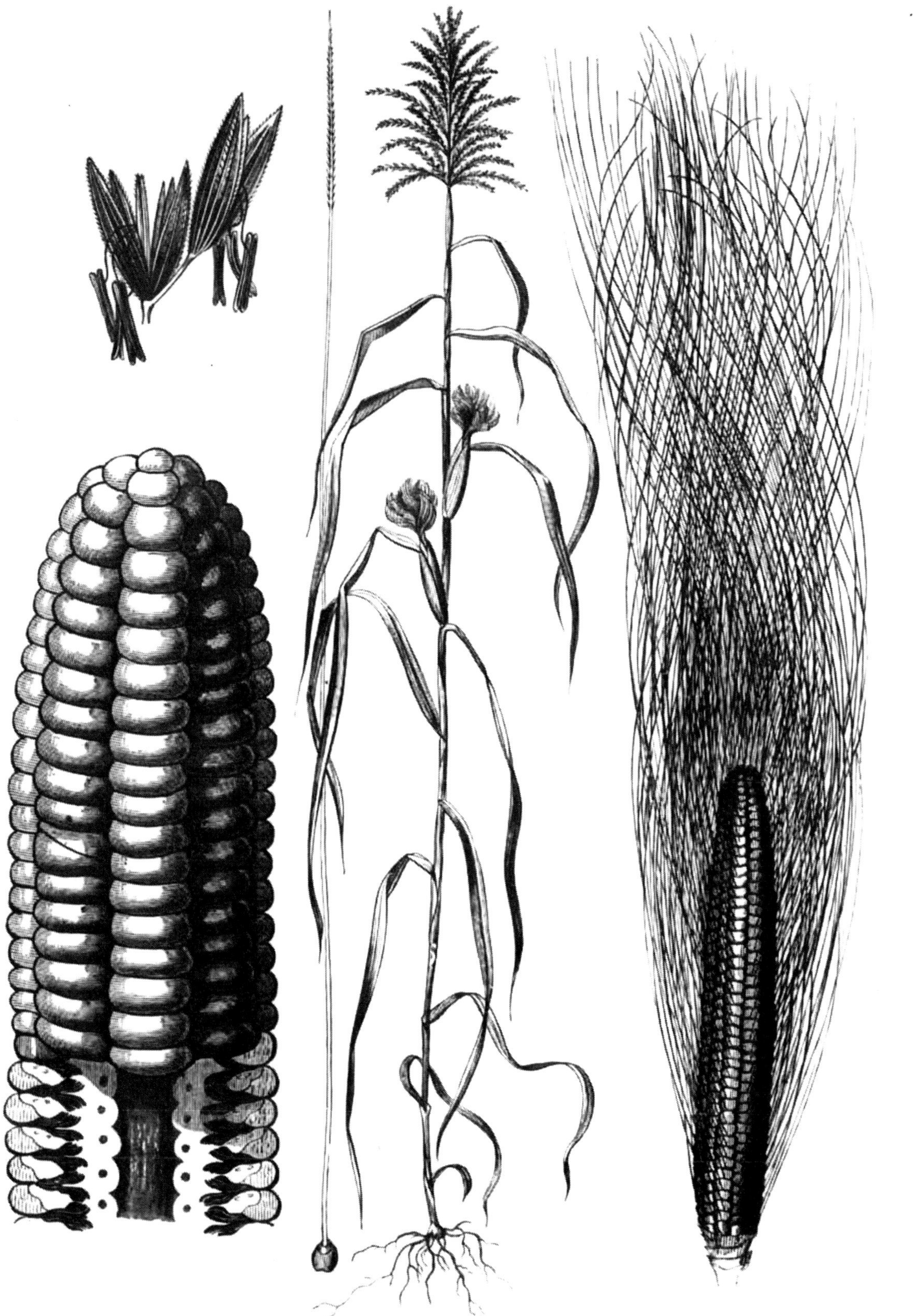

FIG. 11. MAIZE (*Zea Mays*).

MAIZE.

Zea Maÿs, L.

Hind.—Mukka-Bhuta, Bút, Makai, Janara. *Beng.*—Butta, Makka-janar. *Punjab*—Makkai, Kukri, Bari-joar.
Oudh—Bari-jowar. *Tamil*—Makka. *Tel.*—Zonalu.
Sanskrit—Yavanala.

This handsome annual erect grass bears its grain in a solitary axillary female spike. The grain varies much in colour, form, and size.

The maize, though a native of South America, has been long cultivated in many parts of India: there are now over 2¼ million acres under this crop. It is sown in June and July and cut in September.

COMPOSITION OF MAIZE (C.).

	In 100 parts.	In 1 lb.
Water	12·5	2 oz. 0 grs.
Albuminoids	9·5	1 „ 227 „
Starch	70·7	11 „ 137 „
Oil	3·6	0 „ 252 „
Fibre	2·0	0 „ 140 „
Ash	1·7	0 „ 119 „

The nutrient-ratio is here 1 : 8·3, and the nutrient-value 88½.

The above percentages are the means deduced from several analyses of whole Indian-grown maize. They show a lower proportion of water and of oil than the average of European and American samples. The range in the latter, when a few exceptional and clearly abnormal results are excluded, is not wide. Generally, the American-grown maize contains about 1 per cent. more fat or oil than the East Indian.

Maize is extensively used throughout the Benares, Patná, and Bhágulpur Divisions, chiefly amongst the peasantry. Young and green it is roasted and greedily eaten by natives of Gorakhpur, Behar, and Patná. Maize is not considered so wholesome as wheat, being thought rather heating. The stems, both green and dry, and the leaves, are used as cattle food.

The grains are separated from the maize cobs by thrashing. There are three ways of preparing them for food:

a. The grains are roughly ground and made into a kind of porridge or pudding.

b. The grains are parched, pounded, winnowed, sifted, and ground into meal in hand-mills. The dough is made into cakes.

c. The grains, parched in hot sand, become "pop-corn"—chebana, bhuna, kori. This is eaten with gúr or salt by labourers and travellers at their midday meal.

In the United States of North America, and in Peru and Chili, many varieties of maize are in cultivation. Some of the best kinds have large fruits with thin skins and are very prolific, the heads of corn or "cobs" being of considerable size. The pale or white-grained sorts are generally superior to those which have much colour. Esteemed American varieties are—Minnesota Early, Crosby Early, Eight-Row Early, Concord, and Stowell's Evergreen. It is very desirable that more attention should be paid to the selection of the best varieties of maize for cultivation in India.

RICE.

Oryza sativa, L.

Hind.—Beng.—Dhan (cleaned rice is Chauwal, Chawal, Chaol, in *Hind.*). *Tamil*—Arisi. *Telugu*—Úri, Cheni, Matta-Karulu. *Sind*—Sari. *Sinhalese*—Goyang. *Sanskrit*—Vrihi, Arunya, Dhanya.

Rice is an annual grass belonging to the tribe Oryzeæ of the natural order Gramineæ. It grows from 2 to 10 or more feet in height; the panicles vary from 8 inches to a foot or even more in length, and become drooping; the fruit or grain is enclosed in but does not adhere to the pales.

The several rice-crops of India may be termed spring, summer, autumn, and winter-rice, from the seasons in which the different varieties are harvested. Winter-rice is the most important, constituting as it does about three-fourths of the entire amount. Only in Purí, Maldah, Rájsháhí, and Sylhet does spring-rice attain even so high a percentage of the entire crop as 12

67 FIG. 12. RICE (*Oryza sativa*). F 2

to 25; in other localities it is not grown at all or its amount is quite insignificant. The autumn or intermediate crop of rice is likewise of little or no importance, save in about a dozen localities out of 63 concerning which we possess statistics. However, in Patná, Hazáribágh, Purí, and Cuttack, it may reach one-third of the total out-turn, while it amounts to about one-fourth or one-fifth in Santál Parganás, Bánkurá, and Midnapur. Summer-rice is a very general crop throughout India. In one locality, Nuddea, it yields two-thirds of the total amount grown; in Champáran, Bírbhúm, Dacca, and Faridpur, about one-third; and in many other localities, about one-fourth.

The several rice-crops bear different names in different parts of India; in the present section we have generally employed the terms used in Bengal, *Boro* for spring-rice, *Áus* for summer-rice, *Kartika* for autumn or intermediate rice, and *Áman* for winter-rice.

Spring-rice is sown, according to locality, from September to February, and reaped from March to June. Summer-rice is sown from May to July, and reaped from September to October. Autumn-rice is sown in Bengal from April to July, and reaped from August to November; in Jessur it is sown in October and November, and reaped eleven months after. Winter-rice is sown from March to August, and reaped from November to January. Where one crop only of rice is grown in the year it is usually sown from May to August, and reaped from September to January. Where two crops are raised the yield of grain from both crops is little larger than that from one, but the straw of the crop gathered in the dry season, though a wretched fodder, is used for cattle-food. The two chief varieties of rice, winter and summer, are occasionally sown mixed together; sometimes with Panicum miliaceum and Phaseolus Mungo. Peas, oil-seeds, barley, etc., are also largely sown over the nearly ripe crop, which, however, is cut before they appear above ground.

With regard to the several rice-crops, Mr. G. Watt says: "A proprietor of an estate, with a fairly mixed soil, might have three, if not four, or even five, harvests of rice every twelve months, thus:

"(1) *Áus harvest*, from July to August.

"(2) *Chotan áman*, from October to November.

"(3) *Boran áman*, from December to January.

"(4) *Boro*, from April to May.

"(5) *Raida*, from September to October.

"Two harvests are all but universal in Bengal, with an occasional third but smaller one; two crops are frequently taken off the same field." "The long, thin *chotan áman* rices are eaten by the richer natives."

Rice grows well in stiff clays, especially in drainage-beds and basins. Manure is not often used. It is sown in a moist soil, or even in an actual mud, either broadcast or transplanted from a nursery when the plants are something less than a foot high; the distance between the plants is about 6 inches. The yield of transplanted rice is 16 maunds of paddy per acre; when sown broadcast it yields from 10 to 12 maunds. Mr. Duthie states that there are at least 100 cultivated varieties of rice in the North-West Provinces and Oudh; a distinct indigenous species of another genus, Hygrorhiza aristata (Nees), growing wild round lakes and marshes, is gathered and eaten by the poorer classes. The more important varieties of rice are semi-aquatic, and need copious and repeated irrigations. In some districts of Bengal a long-stemmed variety of rice is grown which will keep its head above 12 feet of water. On the other hand, there are varieties of rice which develop in temperate climates, even ascending the hills to an altitude of at least 8,000 feet, and requiring no irrigation.

The analyses which have been made of a large number of samples of "cleaned" rice, give figures which are wonderfully accordant, considering the great differences in the appearance of the specimens and the very diverse conditions under which they have been grown. The fibre and adventitious earth are sometimes rather high from imperfect cleaning of the grain, but the nitrogenous constituents or albuminoids oscillate within

FIG. 13. BEARDED PADDY (*Oryza sativa*).

narrow limits—probably nine samples out of ten will be found to contain not less than 7 per cent., and not more than 8.

COMPOSITION OF RICE.

	In 100 parts.	In 1 lb.	
Water	12·8 ...	2 oz.	21 grs.
Albuminoids	7·3 ...	1 „	74 „
Starch	78·3 ...	12 „	231 „
Oil...	·6 ...	0 „	42 „
Fibre	·4 ...	0 „	28 „
Ash	·6 ...	0 „	42 „

The nutrient-ratio is 1 : 10·8 and the nutrient-value 86½. One hundred parts of rice contain no more than ·065 part of potash and ·284 part of phosphoric acid.

There are many districts in India where rice forms not merely the chief food-stuff but ¾ths or even ⅘ths of its total amount. In some places it even rises to ⅞ths or to 15/16ths of the whole quantity, as in Bardwân, Dinajpur, Maldah, Kuch Behar, Mánbhúm, and Darrang; other districts might be named in which it constitutes the only food staple.

Dhan is rice in the husk, or paddy. *Chaol* is rice husked by pounding in a wooden mortar; in some districts it is, if new, parboiled and then dried before being pounded. Eight pounds of dhan produce 5 pounds of chaol; the separated pericarp is burned, the perisperm is given to fowls and pigs. The operation of pounding is attended with considerable loss, because many grains are broken and then afterwards winnowed away when tossing the rice in the air from the woven straw scoop. *Bhat* is boiled rice. In Tirhút and Sáran the chaol is first washed and then boiled at night, for the evening meal, in much water. It is strained when hot, one-half or one-third being set aside under water (to save the cost of more fuel) for the morning meal—it has then become slightly acidulous. This preparation is eaten with curds, chillies, or one-fourth of dhal (pulse husked and split). Two pounds of cleaned rice weigh 5 pounds after boiling. The liquor is either thrown away or is drunk as a beverage after the addition of a little common salt, or is given to stall-fed milch cows. Where rice constitutes the almost entire food of the population, the throwing away of the water in which it has been boiled involves the loss of some of the mineral matter in which rice is notoriously deficient, and is to be deprecated; no more water should be used in cooking this grain than can be absorbed by it. Rice is sometimes boiled in milk. The parching of rice is often done by stirring it in hot sand and then sifting out the grains. They burst, and are eaten dry, or else are ground, mixed with water, and consumed at midday meals by travellers and labourers. In Maldah the Hindustani-speaking population use rice and wheat, the pure Bengali confines himself to rice. In Benares rice is not much used, being replaced, amongst the poor, by wheat, barley, jowari, bájra, and maize. The industrial and labouring classes of Mirzápur consume but little rice, living chiefly on barley and the various millets.

According to the "Report of the Famine Commission" the percentages of the rice-eating population in 7 provinces, etc., were:

Madras	32	**North-West Provinces and Oudh**	19
Central Provinces	31	**Punjab**	5
Mysore	20	**Berar**	2
Bombay	12		

The exports of rice and paddy from India amounted in 1882–83 to 31¼ million cwts. Sixty-eight per cent. of this came from Burma, 26 per cent. from Bengal, and 4 per cent. from Madras. Of the total exports, 61 per cent. were sent to England or the Continent (including Egypt).

Rice is eaten in many forms and prepared in many ways besides those already described. The five following preparations may be selected for notice:

(1.) *Churwa*, *Chura*, or *Chira*. Some dhan is boiled, dried, and pounded to separate the husks; the chaol thus obtained is then heated in a wide-mouthed earthen pot, and while still hot is flattened by beating. This preparation may be eaten alone, but it is often made into balls with gúr or molasses, or taken with curdled milk (*doyi*), with milk and tamarinds, or with sweetmeats.

(2.) *Alochira* is made by steeping the rough dhan for a night in cold water; it is then parched and afterwards flattened by beating.

(3.) *Khoyi* is made by parching rice which has been exposed to the dew. It is eaten with molasses, constituting *murki*, or with milk.

(4.) *Muri* or *Murhi* is prepared by first heating chaol with salt for about half-an-hour in a shallow earthen vessel kept agitated, and finally parching it. It is eaten by the poor, generally by itself but sometimes with oil.

(5.) *Chaol-ka-atta* is rice-meal made by slow grinding in heavy hand-mills. It is kneaded with water into balls or cakes (*bhaka*), which are boiled like a pudding, or used as bread.

DIETETICS OF RICE.

New rice is cheaper than old, not merely because the latter is more easily husked, nor because it is drier and therefore contains more nutriment in a given weight, but because it is more wholesome. New rice is almost generally considered unwholesome; it is said to cause swellings of the mouth and throat, dyspepsia, diarrhœa, and fever.

Prisoners obliged to live almost wholly on rice become anæmic, symptoms of land-scurvy supervening. Health rapidly deteriorates under its exclusive use, unless it be eaten in excessive quantity; to this course there are obvious objections. Then, too, there are some varieties of rice the "chits" or embryos of which are so hard that they cannot be digested by the strongest stomach; prolonged boiling in water only partially softens them. As a rule, the rice which is sown on swampy ground, which is not transplanted, and which comes to maturity during or just at the close of the rainy season, is of the nature just described. The grains of this rice being large, cheap, and abundant, are much used by the very poor; their use is frequently followed by the disorders of the alimentary canal, skin, and blood, named above. Local names for such rice are *saru* (the worst), *jagar*, *singra*, and *jasuria*. Much illness is occasioned, especially amongst the poor, through the eating of rice which has been imperfectly "cleaned;" dyspepsia, diarrhœa, dysentery, and anæmia are thus caused. It must always, however, be remembered that the very best rice has two capital defects, being deficient in potash, phosphoric acid, lime, and other mineral matters, as well as in nitrogenous or flesh-forming matters, that is, albuminoids. This latter defect is best shown by the statement that while a perfect food should contain less than 5 parts of starch or its equivalent to 1 part of albuminoid, rice contains rather more than 10 of the former constituent to one of the latter.

SUGAR-CANE.

Saccharum officinarum, L.

Beng.—Ik, Úk, Kushiar, etc.
North-West Provinces and Oudh—Ikh, Ukhari. *Telugu*—Cherukú-bodi.
Sanskrit—Ikshu, Rusala, Púndra, Kanguruku.

This strong cane-stemmed grass grows from 8 to 12 feet high, producing a large feathery plume of flowers. It occurs wild and cultivated throughout tropical and sub-tropical Asia.

The "sets" or cuttings of one season's canes are planted in January or February in furrows to the number of about 20,000 to the acre; irrigation is employed until the rains begin.

The sugar-cane grows well on a good loam or light clay. The land needs frequent ploughing and generally a good deal of manure. It has been stated that any considerable quantity of nitrates in the soil or soil-water is prejudicial to this crop; as a general rule, however, all the larger grasses are greatly benefited by small quantities of the nitrates of soda, of lime, or of potash. This crop practically occupies the ground twelve months; usually it is preceded by a year's fallow. The yield of cut canes is said by Mr. Duthie to vary from 18 to 30 maunds per acre

FIG. 14. SUGAR-CANE (*Saccharum officinarum*).

in the North-West Provinces and Oudh. A large succulent variety is grown chiefly as a sweetmeat.

The sugar-canes, cut when about to flower, contain about 74 per cent. of water. The expressed juice contains a small quantity of albumen, ·1 per cent.; of organic acids, ·1 per cent.; and of mineral matter, 1·6 per cent.; but its chief constituent is cane-sugar, accompanied by a small quantity of uncrystallisable or invert-sugar—the latter compound increases greatly if the canes be damaged or bruised and are not immediately pressed or exhausted; traces of another sugar (called raffinose and melitose) are also found in sugar-cane juice and in raw cane-sugar. The total amount of saccharine bodies in the juice of Indian canes approaches 24 per cent.

Owing to defects in the native methods of extracting sugar, much loss and much deterioration of the crystalline sugar present occurs, while the product is impure, dark in colour, and mixed with much molasses. Analyses of samples of native cane-sugar reveal the presence of large quantities of uncrystallisable sugar, moisture, and mineral impurities.

Great Millet or Guinea Corn.

Sorghum vulgare, Pers.

Synonyms—Holcus Sorghum (L.); Holcus bicolor (L.); Andropogon Sorghum (Brot.).

Hind.—Joár, Jawári, Janera, Júndri. *Beng.*—Jowari. *Punjab*—Jawár. *Tamil*—Cholum. *Telugu*—Jonna, Tella-Janular.

Sanskrit—Zúrna.

The culms of this millet are erect, the panicles branched and the grain enclosed in, but free from the hard shining outer glumes. Generally the seed is sown from June or July to September, and the crop cut in October or November up to January. Three to six seers of seed are sown at the beginning of the rains in rather elevated lands of a loamy or clayey kind. Sometimes the minor pulses are sown with it. The yield of grain amounts to 10 maunds, with 60 maunds of stems as fodder if irrigation has been used. Without irrigation, the yield is one-fifth less. To this yield must be added that of the accompanying crop of pulse. Sometimes, as in the North-West Provinces, Oudh, Punjab, joár is grown as cattle-fodder, having been sown and irrigated before the rains and cut green, early enough to make room for the succeeding cold-weather crop. It is very extensively grown in Madras; in 1875–76, more than 4½ million acres were under cholum. There are several well-marked varieties of this Sorghum; they differ chiefly in the form of the panicle, which is sometimes dense and contracted, sometimes made up of drooping branches, and sometimes has expanding branches.

Composition of Great Millet.

	In 100 parts.	In 1 lb.
Water	12·5	2 oz. 0 grs.
Albuminoids	9·3	1 „ 214 „
Starch	72·3	11 „ 248 „
Oil	2·0	0 „ 140 „
Fibre	2·2	0 „ 154 „
Ash	1·7	0 „ 119 „

The nutrient-ratio is here 1 : 8¼, and the nutrient-value 86.

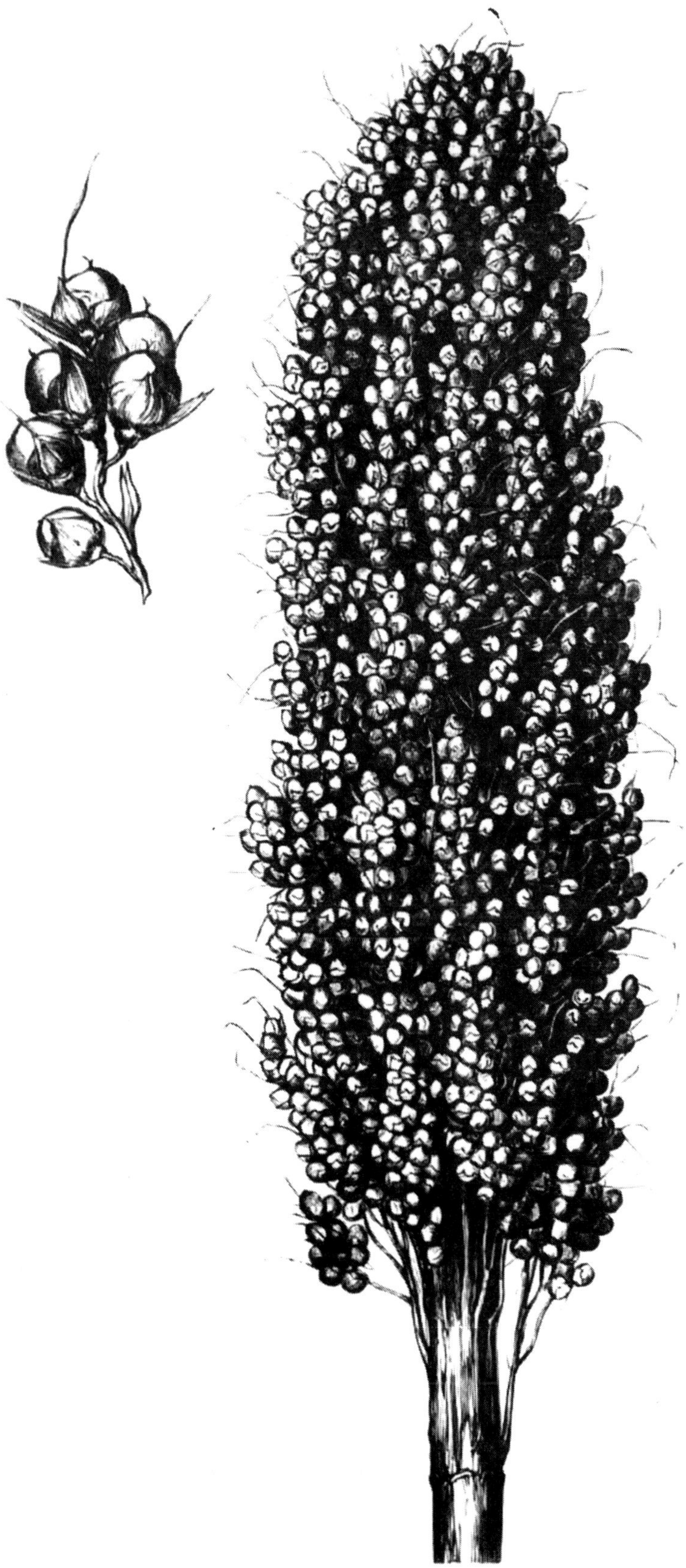

FIG. 15. GREAT MILLET (*Sorghum vulgare*).

FIG. 16. GREAT MILLET (*Sorghum vulgare*).

The grain contains ·85 per cent. phosphoric acid and ·21 per cent. potash.

Joár is one of the most important rainy-season crops of India, forming with rice and wheat the chief staple foods of the country, especially in the Upper Provinces and Bengal. It is more palatable but less wholesome than maize. The meal is made with cakes, or the grains are parched and eaten with salt, or made into a paste and mixed with chillies, gúr, etc.

BROOM CORN.

Sorghum saccharatum, Moench.

Synonyms—Holcus saccharatus (L.); Andropogon saccharatus (Roxb.).
Hind.—Deodhan. *Deccan*—Shalú.

This annual grass is cultivated in some parts of Northern India, either as fodder or on account of the sugar which can be extracted from the stems; the grain, however, is sometimes used as food.

COMPOSITION OF BROOM CORN.

	In 100 parts.	In 1 lb.
Water	12·8	2 oz. 21 grs.
Albuminoids	11·8	1 „ 388 „
Starch	68·3	10 „ 406 „
Oil	3·0	0 „ 210 „
Fibre	3·0	0 „ 210 „
Ash	1·1	0 „ 77 „

The nutrient-ratio is here 1 : 6·4, and the nutrient-value 87.

The stems of this grass contain a good deal of sugar. When young and very immature invert-sugar is present in largest proportion, as they become more mature it diminishes, until when the grain is ripe three-fourths of the saccharine matter is cane-sugar. The total sugar varies from 6 to 18 per cent., the water from 80 to 66; 12 per cent. of cane-sugar is a fair average.

This amount is liable to serious reduction if the sugar be not extracted directly the stems are cut. Some rich canes were found after three weeks to contain no less than 14·7 per cent. of invert-sugar and only 3·6 of cane-sugar. Mechanical injuries to the stems or the slightest touch of frost previous to their becoming ripe, are the chief causes of this deterioration. But the degradation of the cane-sugar occurs very largely during the operation of boiling down the juice, while an immense proportion is lost in the pressed residue.

This grass is used as cattle fodder.

Sorghum halepense (Pers.), "Baru," and S. verticillata (Beauv.). "Chirchitta," yield a grain which, according to Mr. Duthie, is sometimes eaten.

OATS.

Avena sativa, L.

Hind.—Welayti-jow.

This annual grass has been cultivated to some extent of late years in a few parts of India. It is treated much in the same way as barley.

COMPOSITION OF OATS.

	In 100 parts.
Water	12·7
Albuminoids	10·1
Starch	56·0
Oil	2·3
Fibre	16·6
Ash	2·3

The samples from which the above analytical results were obtained do not compare favourably with oats of European growth. The average percentages in these are, 12 albuminoids, 6 oil, 11 fibre, and 3 ash. Sometimes the percentage of albuminoids rises to 15½ and that of oil to 7. A good sample of fresh Scotch *oatmeal* gave me 16 per cent. albuminoids and 10 per cent. oil, with no more than 5 per cent. of moisture; but it must be recollected that 100 lbs. of oats yield no more than 60 lbs. of oatmeal.

Oats are grown in India chiefly for feeding the horses belonging to Europeans; but the richer natives near the chief centres of English influence are beginning to appreciate the value of oatmeal as human food. So far as chemical composition and the ratio of the nutrients are concerned, oatmeal is an almost perfectly adjusted food.

 FIG. 17. RAGÍ (*Eleusine coracana*).

FIG. 18 RAGÍ (*var.* E. stricta, Roxb.).

RAGÍ.

Eleusine coracana, Gaertn.

Hind.—Natchni, Nachani, Nagli, Mandua, Maruya. *Beng.*—Marúa, Modua. *Oudh*—Mindwa. *Punjab*—Mandal, Chalodra. *Punjab and Bengal Hills*—Koda, Kodom. *Himálaya*—Koda.
Deccan and South India—Ragí. *Tamil*—Kaywur. *Telugu*—Kawaru, Sodi, Ponassa. *Sinhalese*—Puta-tana, Kurakkan.
Sanskrit—Rajika.

This semi-erect to decumbent native grass belongs to the tribe Chlorideæ. It is a fairly productive rainy-weather crop for light soils; it may be grown almost upon stones and gravel. It yields from 5 to 6 maunds of grain per acre upon the hills, 12 to 14 maunds in the plains, if carefully cultivated and weeded. It is the staple grain of the Mysore country; sometimes it is there stored in pits and will keep good for years. It is frequently grown with summer-rice, ripening sooner, and thus affording earlier relief in times of scarcity; the straw is used as fodder.

Eleusine ægyptiaca, a closely-allied species found wild on the road-sides of the Punjab and North-West Provinces, and indeed throughout Upper India, yields a poor unpalatable grain which, in times of scarcity, is occasionally collected and eaten. This species is common in the warmer parts of Ceylon.

COMPOSITION OF RAGÍ.

	In 100 parts. Husked.	In 100 parts. Whole.	In 1 lb.
Water	13·2	12·5	2 oz. 0 grs.
Albuminoids	7·3	5·9	0 „ 413 „
Starch	73·2	74·6	11 „ 409 „
Oil	1·5	0·8	0 „ 56 „
Fibre	2·5	3·6	0 „ 252 „
Ash	2·3	2·6	0 „ 182 „

The nutrient-ratio is here 1 : 13, the nutrient-value 84. The percentage of phosphoric acid in the whole grains is about 0·4.

Generally, this millet is sold at a cheaper rate than any other; in some places, in ordinary seasons, 130 to 140 lbs. of it are procurable for about two shillings, and it is looked upon as a famine food, to which recourse is had only in times of drought or of deficient crops. Ragí is, however, commonly used in some districts, as by the poorer and lower classes of Patná Division, also in Bhágulpur, Santál Parganás, Dinájpur, and Gorakhpur. It is more esteemed than maize by the natives of Patná and Behar, but it is not, in general, a popular food, being considered difficult of digestion, productive of flatulence, and astringent. It is much less esteemed than joár and bájra. In Tirhút, cakes made of ragí-flour are largely eaten. In Darjiling, a fermented liquor is prepared from the grain.

WHEAT.

Triticum vulgare, L.

Synonym—Triticum sativum (Lamarck).
Hind.—Kunak, Giún. *Beng.*—Gom.
Persian—Gundum. Vernacular names include—Khani, Mundi, Muria, Ratta, Seta, Kathia, Jamali, with many others.
Sanskrit—Godhúma, Saman.

Wheat is an annual grass of unknown origin. It belongs to the tribe Hordeæ of the order Gramineæ. There are a very large number of cultivated varieties of this plant, distinguished by the presence or absence of an awn, by the colour of the grain—"red" or "white,"—by the hardness and translucency, or by the softness and opacity of the kernel, etc., etc. Great improvements in the size of the grain and its quality and yield have been effected of late years in England, but very little work in this direction has been yet accomplished in India. Continued artificial selection of the best grain has been the method pursued for this purpose; the purity of the strain should be constantly kept in view.

Generally speaking, wheat is grown in those parts of India where rice does not thrive; nearly all the production is included in the region north of the river Tapti; but rarely is it cultivated anywhere south of the Deccan. The average Indian yield has been estimated at 13 bushels per acre, which, considering the very small depth of soil stirred by the light native plough, does not compare unfavourably with the 15 or 16 bushels which is the mean wheat produce in France; in England, however, the average is something like 29 bushels.

The above estimate of 13 bushels per acre (as the average yield of wheat in India) is, we believe, too high. A tabular statement given in the "Report for 1882–83 on the Progress and Condition of India" (page 193) justifies the adoption of a lower average, as will be seen from the figures that follow:

ESTIMATED YIELD OF WHEAT, IN BUSHELS, PER ACRE.

	Land manured and irrigated.	Land fairly cultivated.	Land badly cultivated.
North-West Provinces and Oudh ...	22	15	9
Punjab	20	11	7½
Central Provinces	16	10	6
Bombay	18	10	6
Berar	12	8	5

But the returns on which the above numbers are founded are confessedly imperfect; while in several provinces, and, of course, in most Native States where wheat is grown, the statistics of average yield of dressed corn are incomplete, untrustworthy, or even altogether wanting.

As to the Indian areas under wheat, rough estimates, which are probably not very far from the truth, have been published from time to time. The figures given below are probably within the mark, especially those for the Native States. The chief wheat-growing localities are alone included in the table; in Madras, Assam, and Burma, wheat is not grown to an appreciable extent.

ESTIMATED AREA UNDER WHEAT.

	Acres.	
British India:		
Bengal	850,000	20,350,000
North-West Provinces and Oudh	6,200,000	
Punjab	7,000,000	
Central Provinces	4,000,000	
Bombay	1,600,000	
Berar	700,000	
Native States:		
Hyderabad	750,000	5,838,000
Central India Agency	2,500,000	
Rajputana	2,500,000	
Baroda	88,000	
Grand total ...		26,188,000

Dr. W. W. Hunter estimates the percentages of the Indian food-grain areas under wheat in several provinces to be—

North-West Provinces ...	57	Sind	12
Punjab	54	Bombay	7
Central Provinces	27		

The Indian area of wheat cultivation is defined by Dr. Forbes Watson, in his Report of 1879, as comprising the whole of Northern India up to the Gangetic Delta, and the whole of the table-land above the Ghâts in Southern India. It extends through every district of the North-West Provinces, Oudh, the Punjab, Sind, the Central Provinces, and Berar; also through every part of the Bombay Presidency, with the exception of some of the coast districts; this crop is also grown in many parts of the interior of the Madras Presidency.

Mr. J. F. Duthie, of Sahárunpur, reports that wheat ("gehun" or "gohun") is the largest of all the crops grown in the North-West Provinces and Oudh. It is cultivated on all soils save the lightest sands, although a rather heavy loam suits it best. Many ploughings and a fine tilth are essential; about four tons of dung or village manure per acre are usually applied. The seed is sown broadcast and covered by the plough, or it is dropped into the furrows by hand, or by means of a bamboo attached to the plough-stilt. A watering is given before sowing the grain, when the soil is too dry. Wheat is sometimes grown with barley, sometimes with gram. Fifteen maunds is the average yield of wheat grown by itself on irrigated land, 9 maunds on unirrigated land; wherever possible it is irrigated. The weight of straw varies from 1½ to twice as much as the grain. The wheat is of course a cold-weather crop, being grown in the rabi season, between October and April. In the Deccan it is sometimes sown in September and reaped in January. In the Punjab the harvest is as late as May; in Agra, June.

Dr. Forbes Watson estimated the total wheat production of India in 1877 to be 40 million quarters, of which no more than 1·3 million quarters were exported. The exports in 1881–82 had risen to 4·5 million quarters, but fell in 1882–83 to 3·2 millions. In 1881–82 the value of the wheat exported from India exceeded that of the rice, and amounted to no less than 11 per cent. of the total exports. Forty-six per cent. of the exported wheat (in 1882–83) was destined for England, 25 per cent. for France, and 11 per cent. for Belgium. The fluctuations of the Indian export wheat-trade depend not so much upon the local harvests as upon the yield in the United States and the prices ruling in Europe.

The composition of wheat-grain shows some variations, but they are almost entirely limited to the relative proportions of starch and of nitrogenous matters, although the mineral matters or ash, and indeed all the minor constituents of the grain are, of course, not quite fixed in amount. Still, if a wet season increases the percentage of ash, if a thin-skinned well-developed sample contains less fibre, and if a plump dark-coloured specimen has a larger proportion of oil or fat, all such variations are quite unimportant in comparison with those exhibited by the starch and albuminoids. The starch, always constituting, as it does, something like two-thirds of the weight of the grain, does not show this difference in so marked a manner as the albuminoids. If the latter amount to 18 or 20 per cent., instead of 13, the former constituent will not be reduced (from 68) to less than 63 or 61 per cent.—a reduction which, in comparison with the total amount present, is much less conspicuous than a rise from 13 to 20 in the nitrogenous compounds.

Besides the general dryness of the grain of Indian wheat, which, as imported in bulk into this country and analysed promptly, contains at least 2 per cent. less moisture than average English wheat, the albuminoids are decidedly higher. I have never yet met with an Indian wheat containing less than 10 per cent. of albuminoids; but a large number of samples of first-rate English, Canadian, and Australian samples give numbers between 8 and 9. The average percentage of albuminoids in the Indian examples yet analysed is about 13·5, but some specimens have been as low as 10·3, and some as high as 16·7.

All these estimations are based upon actual determinations of nitrogen, made either by Dumas's absolute nitrogen method or by the easier process of combustion with soda-lime, the percentage of nitrogen obtained being then multiplied by 6·3. Many higher figures have been published, but these, so far as I can learn, have been obtained by a most misleading and incorrect method, namely, the weighing of the so-called "gluten" which is left after washing out the starch from a prepared dough by means of water. The gluten so separated is often largely contaminated with starch as well as with wheat-oil and wheat-fibre, while such contamination is too variable to admit of being allowed for. It is greatly to be regretted that many works on food-materials and dietetics should be disfigured by erroneous figures arising from the use of this unsafe process. As to the maximum percentage of albuminoids present in Indian wheat it is probable that it really does sometimes exceed 16·7, the above-given number, but as yet I have no actual determinations confirmatory of this view.

Much of the Indian wheat, whether white or red, has that translucent aspect which generally indicates a high percentage of albuminoids. In 1867 ("Practice with Science," i., pp. 101–111, 345–348), I pointed out some of the chief relationships between the aspect, density, weight per bushel, productiveness, and chemical composition of wheat-grain, showing more particularly that the exclusion of a small proportion of the lightest of the grains in a seed-corn tells very favourably upon the yield, and also that there is a very intimate connection between the translucency or horny character of a grain and a high percentage of albuminoids; and, again, between the softness and opacity of a grain and a high percentage of starch. Such differences in the composition of wheat-grain show themselves not merely in different varieties of wheat, but even in the same variety of wheat when it has been grown under different conditions of climate or season. Even in the grains from a single ear similar differences may often be seen—analysis showing sometimes 3 or 4 per cent. more albuminoids in some of such grains than in others Often a single grain will be partly horny and partly opaque and soft—in that case its composition will correspond with its intermediate aspect. By examining the cut surfaces of a grain which has been cut transversely with a sharp knife a fair notion of its richness or poverty in albuminoids may indeed be easily gained.

AVERAGE COMPOSITION OF INDIAN WHEAT.

	In 100 parts.	In 1 lb.
Water	12·5 ...	2 oz. 0 grs.
Albuminoids	13·5 ...	2 ,, 70 ,,
Starch	68·4 ...	10 ,, 413 ,,
Oil	1·2 ...	0 ,, 84 ,,
Fibre	2·7 ...	0 ,, 189 ,,
Ash	1·7 ...	0 ,, 119 ,,

The nutrient-ratio is here 1 : 5·2, and the nutrient-value 84·6. It should be added that the starch named above contains a small quantity, about 2 per cent., of the sugar or sugars found in many cereals, but this may be regarded as not appreciably lowering the nutrient-value of the 68·4 parts set down as "starch."

The ash of wheat, though not large in amount, is of great importance as a source of mineral nutrients when this grain is used as human food, about 30 per cent. of it being potash and 45 per cent. phosphoric acid.

The various mill-products obtained in grinding wheat differ much from each other and from the original grain, in several important particulars. For instance, the following figures were obtained in a series of analyses which I made of an entire series of such mill-products:

	Per cent. Nitrogen.	Per cent. Oil.
Whole Wheat	1·692 ...	2·02
Flour (whites)	1·621 ...	1·4
Flour (seconds)	1·967 ...	1·82
Bran	2·143 ...	2·75
Sharps (fine)	2·608 ...	3·50

These products represent but four out of a total of twelve, but they suffice to show how large a proportion of nitrogenous matter and of oil are rejected when fine flour is the sole product reserved for human food. It must not, however, be assumed that all the nitrogen, say, in fine sharps, is albuminoid; in fact, these fine sharps did not contain more than 13½ per cent. of albuminoids, though 2·608 of nitrogen corresponds to 16½ per cent.; even the fine "whites" or flour contained a little nitrogen in non-albuminoid forms.

The methods of employing wheat for human food in India vary somewhat, but the following are used to a considerable extent. The grain having been separated from the chaff, often rather imperfectly, is washed, sun-dried, and then ground between millstones into meal. The finest part or sují, the second grade or maida, and the coarsest or atta, are respectively used as follows:

Sují and maida are employed chiefly in the making of confectionery, while the atta is made into unleavened bread or biscuits, usually in the form of flat cakes called chapattí or rotí. These cakes constitute one of the chief articles of diet in many parts of India, as in Monghyr, Gorakhpur, and Behar; they are eaten with dal, ghí, etc. They are prepared by kneading the flour with water into a dough, which is pressed into cakes and baked over a fire or on a hot earthen platter. Fried with ghí and sugar they are largely consumed by rich Hindus. A mixture of wheaten and barley flour is employed in some districts for making the chapattí. Fermented bread is, generally speaking, unknown in India, but it is eaten, both by Hindus and Moslems, especially by the rich and middle classes, in the principal towns of Patná and Behar. It should be mentioned here that scorbutic affections do not occur where wheat is a considerable or almost exclusive article of the daily dietary, a fact in marked contrast with the results observed in districts where rice is very largely consumed.

A careful study and practical examination of the milling qualities of certain typical samples of Indian wheat was made by Messrs. McDougall Brothers, of Mark Lane, in 1882. From their report, addressed to the Secretary of State for India, we tabulate the following particulars as to the four samples with which the experiments were made:

	Value (496 lbs.)	Weight per bush.	Weight of 100 grs.
No. 1. Fine soft white	49s.	64 lbs.	55·4
No. 2. Superior soft red	45s.	62¾ "	51·8
No. 3. Average hard white	44s.	60 "	68·3
No. 4. Average hard red	43s.	61¼ "	77·7

		Per cent. Flour.	Middlings.	Pollard.	Bran.
No. 1.	Ground under millstones	77·46	·82	8·8	12·0
	Crushed between rollers	74·1	11·0	8·7	4·0
No. 2.	Ground under millstones	78·0	1·68	9·8	9·4
	Crushed between rollers	75·4	7·7	13·5	5·5
No. 3.	Ground under millstones	80·52	·78	10·0	8·3
	Crushed between rollers	73·2	10·3	14·3	3·1
No. 4.	Ground under millstones	79·88	·78	13·2	8·5
	Crushed between rollers	74·2	10·3	13·8	3·0

No. 1.—Is a wheat of medium-sized grain and high weight per bushel. One-and-a-half per cent. of impurities were removed before grinding it, and it was allowed to absorb 2 per cent. of water. It yielded a great quantity of pure white flour of superior bloom and having a slight beany flavour.

No. 2.—Half the grains were medium-sized, half small. ·72 per cent. of impurities were removed before grinding it, and it absorbed 3·6 per cent. of water. It yielded a great quantity of flour of medium colour and strength.

No. 3.—Grains translucent and long. 3·7 per cent. of impurities were removed before grinding; being very dry it absorbed 8·4 per cent. of water—it and No. 4 probably unsurpassed for quantity of flour; the colour of the flour from No. 3 was medium.

No. 4.—Grains translucent, large, long, and coarse. 1·2 per cent. of dirt, etc., were removed before grinding it, and it absorbed 7·6 per cent. of water. The yield of flour (which was of medium colour with good bloom) was remarkably high.

These Indian wheats require a liberal admixture of English or American, or "cold-country" wheat. The flours have an aromatic or beany flavour, and need skill in mixing the grists. Probably from 25 to 50 per cent. of American wheats having a sweet, milky, or nutty flavour would be required. The flours are "ricey," and yield a very large quantity of bread. The bread is too close and the crusts too hard and brittle when flours from unmixed Indian wheats are employed.

The value of Indian wheats in European markets is often much lowered by preventable impurities. Very frequently they contain other cereal grains, especially barley; gram and linseed sometimes occur in them, and they are often largely contaminated with sand and earth. Then, also, two or more kinds of wheat are found mixed together—hard wheat with soft, and red wheat with white.

While the soft wheats abundantly produced in Northern India are better adapted for milling purposes, the hard wheats of Southern India will prove particularly appropriate for the manufacture of macaroni, vermicelli, and *pâtes d'Italie*. This is owing to the very high percentage of gluten which they contain—a point in which they resemble the Polish and other European wheats so much prized by the makers of macaroni.

It appears that hard wheats, both white and red, are preferred for local consumption in India, and formerly commanded the highest prices. Hard white wheat is grown chiefly in the Deccan and Southern Mahratta country; hard red wheat chiefly in Rájputána, the Central Provinces, and Bombay generally. Soft white wheat is grown to great perfection throughout Northern India, and also in Rájputána and Gujarát. It prefers a rich loam, well manured and irrigated, and a moderately severe winter. The most valuable sort of soft red wheat, under the name of pissi, comes from the Central Provinces, and especially from the Narbada valley. The predominant soil in this tract is a heavy black loam—"black cotton soil"—which retains moisture throughout the year, and consequently needs no irrigation.

The straw of wheat is sometimes used in India as fodder, alone or mixed with barley straw and the haulms of pulse.

BARLEY.

Hordeum vulgare, L.

Hind.—Jau, Jow, Jawa. *Beng.*—Jab. *Tamil*—Barali-arishi. *Telugu*—Yavala, Barali-bujyam. *Deccan*—Satú. *Guj.*—Jau, Ymwah. *Burma*—Muya. *Nepál*—Tosa. *Kangra*—Yoa. *North-West Provinces*—Indurjou, Yurk. *Sanskrit*—Situshúka, Yuva.

This well-known annual grass produces many stems, 2 to 3 feet high, from a single grain. There are many strongly marked forms of this cereal, but it is now generally believed by botanists that they are all referable to a single species. The variety called "Six-rowed" (*H. hexastichon*) is the one usually grown in India; another kind, cultivated in the inner Himálaya and Tibet, much resembles wheat in the form of the ear, and in the fact that the grains are naked.

The cultivation of barley is practically confined to Northern India, rather more than one-third of the total being grown in the Punjab, and a little under two-thirds in the North-West; the Allahabad and Benares Divisions show the largest areas in the latter province. In the Punjab this crop is heaviest in the South and South-East, and scarcely extends along or across the Indus. In the north of Bombay a little barley is grown.

Generally barley is sown in October–November, and reaped in February, March, or April. Where indigo is grown in the kharif, barley is its usual accompaniment in the rabi. It is often cultivated in light or sandy soils and usually receives but little manure. Messrs. Duthie and Fuller estimated the yield of grain when it is grown alone at 16 maunds when twice irrigated, at 10 maunds when not irrigated; the straw amounts to about 1½ times the weight of the grain. It is sown and cultivated in the same way as wheat, but needs fewer ploughings. It is often grown with wheat or pulse.

COMPOSITION OF BARLEY (HUSKED).

	In 100 parts.	In 1 lb.
Water	12·5 ...	2 oz. 0 grs.
Albuminoids	11·5 ...	1 „ 368 „
Starch	70·0 ...	11 „ 87 „
Fat...	1·3 ...	0 „ 91 „
Fibre	2·6 ...	0 „ 182 „
Ash	2·1 ...	0 „ 147 „

The nutrient-ratio is here 1 : 6·3, and the nutrient-value 84·5.

The above analysis of a roughly-cleaned or husked sample of Indian barley, shows that it contains a higher percentage of albuminoids than average European barleys. This fact is confirmed by the examination of other Indian samples. At the same time it must be recollected that it is inferior to Indian wheat in this particular, just as the hundreds of analyses which have now been made of European barley show that it likewise gives a lower average percentage of albuminoids than European wheat. When barley is completely cleaned or pearled, it loses a very large proportion of its albuminoids, so that European pearl barley (barley yields but 38 per cent. of pearl barley) does not usually show more than 6 or 7 per cent. of albuminoids. The "pearl dust" and "fine dust" separated in its preparation, and amounting together to 40 per cent. of the original grain, are however much richer, containing 12 or 14 per cent. of albuminoids; Indian pearl barley would, however, in all probability contain as much as 10 per cent.

Barley, as it is prepared for human food in India, is generally considered to be rather difficult of digestion. It is grown and eaten throughout the whole of the Patná Division. With wheat it forms an important staple diet in the Benares and surrounding divisions. The grain is usually cleaned by pounding in wooden mortars and winnowing. The grain is treated in one or other of the following ways:

(1.) Ground into coarse meal and made into chapatti either alone or with wheat meal. In Tirhút a mixture is used of barley 1 part, Indian corn 3 parts.

(2.) Parched and ground into coarse flour called suttú; this is stirred up with sufficient water to make a thick paste, to this a little salt is added, and the preparation is eaten with garlic, onions, or chillies. This mixture, generally admixed with flour of gram or other seeds or grains, forms the chief food of the larger part of the peasantry of Shahabad.

Barley alone or even in admixtures is generally thought to be rather difficult of digestion, at least in the form in which the grain is prepared for food in India. Barley mixed with horse-gram forms an excellent food for horses, and is known as "adour."

From an observation made in a preceding paragraph it will have been remarked how very closely the pearl barley prepared in Europe approaches rice in its nutrient-ratio. The Indian cleaned barley is, as we have seen, much richer in albuminoids. This arises from two causes, one of which is the higher percentage of nitrogen naturally present in the average whole barley grain as grown in India; the other is the imperfect way in which the Indian barley is cleaned previous to use as food. Some room there evidently is for improvements in the mode of carrying out the cleaning or pearling operation. A recently invented Dutch process might be used. It produces a pearled grain of larger size than that obtained by the usual operations; the pearled grain attains a higher percentage; its shape is not spherical, but much resembles that of the whole grain, and it is richer than the ordinary sort in oil, mineral matter, and albuminoids. The following details as to the great losses incurred in the English method of pearling barley drawn from my own investigations, may be useful:

100 lbs. of barley yield 12½ lbs. of "Coarse dust," and become "blocked barley."

Blocked barley yields 14¾ lbs. of "Fine dust," and becomes "pot or Scotch barley."

Pot barley yields 25¼ lbs. of "Pearl dust," and becomes "pearl barley."

The quantity of pearl barley thus obtained is about 37⅝ lbs., a loss of 10 per cent. being unaccounted for.

The composition of the three waste products or "dusts" is in 100 parts:

	Coarse Dust.	Fine Dust.	Pearl Dust.
Water	14·2	13·1	13·3
Albuminoids	7·0	17·6	12·1
Oil	1·7	6·0	3·4
Starch	46·9	50·5	67·2
Fibre	24·5	8·5	1·8
Ash	5·7	4·3	2·2

In point of fact the albuminoids in all three analyses are much exaggerated, because the nitrogen from which they are calculated does not all exist in the albuminoid form. As an example of this correction, 6 parts must be deducted from the 17·6 per cent. of analysis 2—even then the 11.6 left is a high amount.

BAMBOO RICE.

Bambusa arundinacea, Willd.

Beng.—Bansh, Bans. *Punjab*—Magar bans, Nal bans. *North-West India*—Kattung. *Bomb.*—Mandgay. *Telugu*—Vedúrú. *Burma*—Ki-a-kap-wa. *Sinhalese*—Kattú-úna-gass. *Tamil*—Mungil. *Sanskrit*—Vausa.

This species of bamboo grows in large compact clumps of 30 to 100 stems, and generally attains a height of 30 to 50 feet, but is much taller on the west coast. The leaves are narrow, lanceolate, and from 4 to 8 inches long. The flowers appear at long intervals, all the stems of one clump being covered with flowers in the same season. The fruit (a caryopsis) much resembles in form and size the common oat, and is enclosed in a glume and palea.

This plant is a native of Southern India, Belgaun, Jubbulpur; it is cultivated in the sub-Himalayan tract of the Punjab. It occurs commonly on the margin of rivers in the warmer parts of Ceylon.

COMPOSITION OF BAMBOO GRAIN (HUSKED).

	In 100 parts.	In 1 lb.
Water	11·0 ...	1 oz. 332 grs.
Albuminoids	11·8 ...	1 „ 388 „
Starch	73·7 ...	11 „ 346 „
Oil	0·6 ...	0 „ 42 „
Fibre	1·7 ...	0 „ 119 „
Ash	1·2 ...	0 „ 84 „

The nutrient-ratio is here 1 : 6·4, and the nutrient-value 87. The food value of bamboo grain, after the removal of the husk, is high; its defects are due to the low proportion of oil and of mineral matter. Of course it cannot be looked upon as a staple cereal, but, as an occasional substitute for a deficient rice or millet crop, it has several times proved most serviceable. General Munro considers that it was this species (B. arundinacea) which, in 1864, during one of its occasional flowerings, furnished food to upwards of 50,000 persons in Kanara. It is stated that in 1812, in Orissa, a general flowering of the bamboo prevented a famine. The grain of other kinds of bamboo is, in all probability, similar in composition to that of B. arundinacea. The three most important species are: B. spinosa (Roxb.), growing in Bengal, Gongachora, Gorághát, Assam, Dacca, Cáchár, Tenasserim, Moulmein, Mergui, Pegu; B. tulda (Roxb.), plains of Bengal and adjoining provinces; and B. vulgaris (Wendl.), which is cultivated in many districts of India, particularly in the Western Deccan, Kolapur, Sattara, Poona, Sylhet, Cáchár, Chittagong, and in the Eastern Punjab; this last species grows also in Ceylon, where it is known as "una."

Young shoots of bamboo are eaten like asparagus; they are sometimes pickled. It is a favourite fodder. The plants, or rather the stems, do not survive the seeding—at least this seems to be generally the case with the four species above named.

General Munro ("Linn. Soc. Trans.," xxvi., pp. 87 to 157) describes 23 species of Bambusa as well as many allied plants belonging to other genera of Bambuseæ. There is evidence that the grain of not a few of these has been used, or may be used, for human food.

PART IV.

BUCKWHEAT AND ITS ALLIES.

THE Cock's Comb, Prince's Feather, and Love-lies-bleeding, the Spinach and Beet, the Rhubarb, Sorrel, and Dock, may be cited as familiar examples of the three types of plants which we have grouped together in the present part of this handbook. These plants are generally of weedy habit and rapid growth, and delight in soils containing much nitrogenous matter. Indeed, the succulent stems of many of the species contain at one time or another of their growth—generally just before flowering—an extraordinary amount of nitrates; in the case of some species of Amarantus nearly 15 per cent. of these salts has been found. The fruits or seeds of these plants do not, so far as we know, widely differ in their chemical composition from the millets considered in the previous part of the present work, but complete analyses of several different kinds are still wanting. However, as might be expected from the extreme richness of the succulent parenchymatous tissues of these plants in nitrates just before they flower and the disappearance of those salts afterwards, the percentage of albuminoids (which are formed from nitrates) in the seeds is high, ranging between 13 and 19. Taken as a whole, this group of food grains shows a nutrient-ratio between the albuminoids and starch which closely approaches that demanded in a perfect food, while the proportion of oil and of useful mineral matter is also quite satisfactory.

It should be noted that the seeds of Celosia, Amarantus, and Chenopodium are not enclosed, as are those of Fagopyrum, in a very thick husk or pericarp.

FIG. 19. COMMON AMARANTH (*Amarantus paniculatus*).

The nutrient-ratio and nutrient-value of the three chief members of the present group will stand thus:

	Nutrient-ratio.	Nutrient-value.
Amaranth	1 : 5·3	... 90
Buckwheat	1 : 4·7	... 86
Quinoa	1 : 3	... 78

The plants yielding the products to be discussed in the present chapter belong to three nearly related Natural Orders:

Order, AMARANTACEÆ.	Genera.
Tribe i. CELOSIEÆ.	Celosia.
ii. AMARANTACEÆ.	Amarantus.
Order, CHENOPODIACEÆ.	
Tribe i. EUCHENOPODIEÆ.	Chenopodium.
Order, POLYGONACEÆ.	
Tribe iii. EUPOLYGONEÆ.	Fagopyrum.

Celosia cristata, L.

Punjab—Sil, Siyal, Síya, Mawal, Dhurra.

An erect herbaceous annual, having smooth, generally oval, leaves. The numerous seeds are lenticular or reniform. It occurs throughout India, cultivated or as an escape.

The seeds of this species of Celosia have not been submitted to detailed chemical analysis, but there are good reasons for concluding that, while generally inferior as a food-grain to Amarantus, it does not differ widely in chemical composition from that of the seeds of the common kind of Amarantus.

COMMON AMARANTH.

Amarantus paniculatus, L.

Synonyms—A. frumentaceus (Ham. in Roxb. "Flora Indica"); A. anardana (Ham. in Wall. Cat.); A. speciosus (Sims); A. sanguineus (Linn.); A. strictus (Willd.).

Beng.—Sag. *Tamil*—Púng-kíray.

Punjab—Ganhar, sil, sinl, sawal, bhabri, savalana, batu, daukar, chaulei, etc.

A tall robust annual, stem striate, leaves long-petioled, elliptic or ovate-lanceolate acute or finely acuminate. Seeds yellowish white or black, with or without a border; they are generally $\frac{1}{20}$th inch in diameter, but variable in size as well as in form and colour. Some of the above-cited synonyms may belong to A. caudatus, a closely allied species.

This plant is cultivated throughout India and Ceylon, and up to 9,000 feet in the Himálaya.

The three analyses given below were made upon samples of seeds respectively identified with: A. frumentaceus (analysis A), A. anardana (analysis B), and a third form of the exact character of which some doubt exists (analysis C); analysis B is a recent one of my own.

COMPOSITION OF AMARANTUS PANICULATUS (SEEDS).

	In 100 parts. Analysis A.	Analysis B.	Analysis C.	In 1 lb. Analysis B.
Water	11·9	12·5	12·6	2 oz. 0 grs.
Albuminoids	13·7	14·3	13·6	2 „ 126 „
Starch	58·4	60·4	55·4	9 „ 290 „
Oil	6·0	6·8	6·4	1 „ 39 „
Fibre	4·8	3·4	6·8	0 „ 238 „
Ash	5·2	2·6	5·2	0 „ 182 „

The nutrient-ratio is 1 : 5·3, and the nutrient-value 90.

The red amaranth grown by the Bodagars on the Nílgirís and described as A. frumentaceus by Buchanan, is regarded by Cleghorn as identical with the "báthu" of the North-West Himálaya, which forms so remarkable a feature in the landscape at Simla in October and November. Dr. Wight says that "it is much cultivated on the slopes of the higher hills in several districts of Southern India. In Coimbatore, Salem, and Madura, I have frequently met with large fields of it, often on very steep slopes; in such situations it often grows upwards of 6 feet high. The seed ground into meal forms the principal food of the wild inhabitants of these hills." Probably this is the same plant as "rajgírah," which, according to Colonel Sykes, is cultivated in the Deccan. He says it is not a bread-grain, but is eaten by those Hindús who keep the two fast days of each month. The practice is to parch the seeds, to reduce them to meal, and to eat this meal mixed with sugar. The seeds are sown during the rains; the crop is ripe from October to February; the green leaves are used as a fresh vegetable. One plant will produce 100,000 seeds.

BENGAL AMARANTH.

Amarantus gangeticus, L.

Synonyms—A. tricolor (L.); A. tristis (L.); A. melancholicus (L.); A. lanceolatus (Roxb.); A. oleraceus (Roxb.); A. polygamus (Roxb.); A. lividus (Roxb.); A. amboinicus (Herb. Ham. in Wall. Cat.); A. inamœnus (Willd.).
Hind.—Rukta-shaka, Lab-shak. *Beng.*—Lalsag, dengúd.

An erect, stout, handsome species, generally 2 to 3 feet high; leaves 2 to 5 inches, long-petioled, ovate-oblong, or lanceolate obtuse, but very variable; seed $\frac{1}{20}$th inch in diameter, lenticular and black.

Cultivated and in cultivated ground throughout India and Ceylon; common in Bengal.

COMPOSITION OF AMARANTUS GANGETICUS (HUSKED SEEDS), C.

	In 100 parts.	In 1 lb.
Water	12·1	1 oz. 410 grs.
Albuminoids	14·5	2 „ 140 „
Starch	61·8	9 „ 388 „
Oil	6·3	1 „ 3 „
Fibre	2·6	0 „ 182 „
Ash	2·6	0 „ 182 „

The nutrient-ratio is here 1 : 5¼, and the nutrient-value nearly 91. The analysis shows that we have in these seeds a food in which the proportions, not merely of albuminoids to total starch plus the starch-equivalent of the oil, but also of the oil itself, are very nearly those of an ideal or standard ratio. The Amaranth last described is almost identical with this species in this important respect.

Chenopodium album, L.

Punjab—Irr, Bathua, Jausag, Lunak, Mustakh, Bajar banj, Ratta, Siriari, Taku, Gaddi-Sinugar.
Ladákh—Em.

This erect herb has rhomb-ovate, toothed, somewhat powdery leaves; the flowers are small and green, the seeds lenticular.

It is cultivated in Bengal, in the Punjab, and at considerable heights on the Himálaya. The same species is said to be grown in Kashmír, largely in the Province of Ladákh.

The leaves of Ch. album are used as a pot-herb and as a green vegetable. They are rich in mineral matters, particularly in potash salts. They likewise contain a considerable amount of albuminoids and of other compounds of nitrogen. The seeds, of which we possess no complete analysis, are considered superior to buckwheat.

QUINOA SEED.

Chenopodium Quinoa, Willd.

This species of Chenopodium was introduced from Peru. Its leaves have long slender petioles; they are ovate, wedge-shaped at the base, and pulverulent, with a glaucous or at length reddish meal; the seeds are shining, with a sub-acute margin. The stem is stout and erect, and sometimes attains a height of 5 feet. The flowers are produced in panicles, both axillary and terminal; they are green and inconspicuous.

A light argillaceous soil or a loam is suitable for the growth of this plant. It may be sown in furrows or on ridges a yard apart, the plants in the rows being singled so as to leave 2 feet between each; but it yields heavier heads of grain when sown in nurseries and then transplanted to the fields. The harvest takes place seven months after seed-time. It is a hardy plant, which may be grown successfully in countries and climates and at elevations above the sea such as are suited to the barley crop.

According to an analysis made by the late Dr. A. Völcker of the bitter variety of Quinoa seed—which variety appears to differ from the ordinary sort only in the presence of a small quantity of a bitter and acrid substance—the following numbers represent the

COMPOSITION OF QUINOA.

	In 100 parts.	In 1 lb.
Water	16·0	2 oz. 245 grs.
Albuminoids	19·2	3 „ 32 „
Starch, etc.	47·8	7 „ 283 „
Oil	4·8	0 „ 336 „
Fibre	8·0	1 „ 122 „
Ash	4·2	0 „ 294 „

FIG 20. QUINOA (*Chenopodium Quinoa*).

The ash is rich in potash and phosphoric acid, containing one-third its weight of each of these constituents.

The nutrient-ratio is 1 : 3, and the nutrient-value 78. Were it not for the rather high proportion of indigestible fibre present in Quinoa seed, it would constitute a food of remarkable richness. As it is, it equals several of the pulses in its percentage of albuminoids, while it contains no insignificant amount of oily matter. There is considerable nutritive resemblance between Quinoa seeds and oats; both products are capable of sustaining life in a healthy condition without the aid of other foods.

Quinoa seeds should be soaked for a short time in water and boiled quickly therein; then this water should be thoroughly drained away on a sieve or colander, and the cooking finally completed in another portion of fresh water. By this treatment the bitter principle present in one of the chief varieties of this seed is removed.

Quinoa seeds are extremely small; 15,000 weigh no more than an ounce.

BUCKWHEAT.

Fagopyrum esculentum, Moench.

Synonyms—F. vulgare (Endl.); F. sarracenicum (Dumort.); Polygonum fagopyrum (L.); P. dioicum (Hamilt.).
Punjab—Kala-Trumba, Chin, Kathu, Bras, Tsubri, Phápra.

An upright annual herb of quick growth; stem, hollow and angular; leaves, hastate or cordate triangular; fruit, trigonous, with keeled edges.

Northern India, ascending to 11,500 feet in Kumaon.

In the absence of any analysis of Indian-grown common buckwheat, the figures that follow may be taken as representing the ordinary composition of the husked seeds of this plant.

COMPOSITION OF BUCKWHEAT (HUSKED), C.

	In 100 parts.	In 1 lb.
Water	13·4 ...	2 oz. 63 grs.
Albuminoids	15·2 ...	2 „ 189 „
Starch	63·6 ...	10 „ 77 „
Oil	3·4 ...	0 „ 238 „
Fibre	2·1 ...	0 „ 147 „
Ash	2·3 ...	0 „ 161 „

The nutrient-ratio is here 1 : 4·7, and the nutrient-value 86.

No analyses have been published of the following species of Fagopyrum grown in India: F. emarginatum, Roth. (Nepál, Kunáwár); F. cymosum, Meisn. (Nepál, Mussorí, Kashmír, Kumáun); F. triangulare, Meisn. (Nepál, Kumáun, Sírmúr, Assam); and F. rotundatum, Bab. (Kunáwár, Kumáun).

Buckwheat is used by the poorer classes in some parts of Upper India as food; the seeds are ground into meal and made into thin cakes.

Fagopyrum tataricum, Gaertn.

Synonyms—F. dentatum (Moench.); Fagotriticum sibiricum (L.); Frumentum sarracenicum (Act. Nat. Cur.); Polygonum tataricum (L.).
North-West Provinces—Daran.

The six species of buckwheat grown in India are so similar that they are generally confounded together by the natives

Fig. 21. Buckwheat (*Fagopyrum esculentum*).

under the same names—"kotu," "bro," etc. None of the buckwheats are considered equal to the millets even in the hills; in the plains they are regarded as heating and unpalatable. They make a bitter, poor, and hard bread. The species are all cultivated to a great height, even up to 14,000 feet in Ladak; the crop is an autumnal one.

An imperfect chemical analysis of the fruits or unhusked seeds of the present species shows it to resemble very closely the common kind cultivated in Europe, the albuminoids being 10·9 per cent., the oil 2·4, and the ash 7; the percentages of albuminoids and oil would be considerably raised by the removal of the husk.

The buckwheats are amongst the foods which may be lawfully eaten on the Hindu fast days.

PART V.

PULSE OR LEGUMINOUS SEEDS.

The Characteristics of Pulse—Preparation of Pulse for Food—Use of Pulse in India—Botanical and Chemical Classification of Pulses.

THE seeds of leguminous plants, generally known as *pulse*, differ chemically from the cereal grains in several particulars. They sometimes contain rather more oil or fat—a constituent which may rise even to 17 per cent. (in Inga-beans), to 18 per cent. (in soy-beans), or to 50 per cent. (in pea-nuts). They rarely yield less than 2½ per cent., and often as much as 4, of mineral matter or ash. More important, however, than either of these constituents, is the nitrogenous matter of pulse. This is often called legumin, or vegetable casein, but in reality it varies in different kinds of pulse, and it is a mixture, not a single definite compound. Three substances have been separated from the so-called vegetable casein, none of them being identical in all respects with the casein of milk. They are called respectively, gluten-casein, legumin, and conglutin, but it would be hazardous to affirm that the substances described under these names are perfectly definite compounds. They contain, in 100 parts—

Carbon ...	... ...	51·	to	54·
Hydrogen	... ...	6·8	,,	7·1
Nitrogen	... ...	14·8	,,	18·5
Oxygen ...	... ...	22·	,,	27·
Sulphur ...	... ...	·5	,,	1·

Legumin occurs in largest proportion and in the larger number of kinds of pulse; conglutin is said to be chiefly characteristic of the lupine. With all these compounds much phosphate or phosphorus is associated; it is always difficult and sometimes impossible to separate this substance from the vegetable caseins.

For the purpose of calculation it will be quite sufficient to assume one composition for the albuminoids of all vegetable foods including pulse. We assume, throughout the present work, that all these albuminoids contain 15·87 per cent. of nitrogen, so that we may always calculate the amount of albuminoids present by multiplying the nitrogen found in an analysis by the coefficient 6·3; for $15{\cdot}87 \times 6{\cdot}3 = 99{\cdot}981$, practically 100. It is quite true that some of the samples of legumin extracted from pulse contain nearly 17 per cent. of nitrogen, but the amount is often lower, and we shall not be led into any serious error by adopting the figure named above.

Another point connected with the nitrogen of pulse must be here noted. Some of this nitrogen exists in the form of nitrogen compounds which are not albuminoid—which are not flesh-formers, in fact, and which, for all we know, may be entirely without nutritive value. These substances are simpler in constitution than the albuminoids, and are often of the nature of alkaloids—lupinine, a bitter basic substance from lupines, is one of these, asparagine is another. But the quantity of nitrogen existing in pulse in the form of non-albuminoid compounds of all kinds is small, not exceeding 3 to 5 per cent. of the total albuminoids in the common kinds of ripe pulse; in the seeds, stems, and pods of the unripe plants it is very much larger.

The digestibility of the albuminoids in pulse as compared with that of the corresponding compounds in the cereal grains, has been usually regarded as low. In general, they are not only digested and absorbed at a slower rate, but a larger proportion of the total amount present remains unattacked and unused in its passage along the alimentary tract. The proportion of unused to used albuminoids is proportionately highest when the pulse forms the largest part of the ration; it is much reduced when the pulse constitutes not more than one-fourth of the daily food; and it is still further lowered when the pulse is eaten with milk, butter and eggs, or with other easily digested animal foods. Even under favourable circumstances the unabsorbed portion amounts to 8 per cent. of the total. Of starch in pulse from 93 to 96½ per cent. may be taken up, but the fat or oil is less available, except in the case of the more oleaginous kinds, such as soy-beans and pea-nuts.

Many kinds of pulse should be prepared for food by first removing the seed-coats or skins of the seeds. The slow but thorough cooking of the meal obtained by grinding the split seeds is important. As illustrations of the various modes of preparing pulse for food, the plans adopted in certain localities for treating chick-peas and pigeon-peas may be cited.

Chick-peas are thus used; they are either—

i. Parched and then eaten with or without oil; or they are parched and ground into coarse meal which is stirred up with water, garlic, onion, or a chillie being added.

ii. Ground into flour and made into chapatties or sweet cakes, these generally, however, containing other flours; or the flour is made into balls with water and spices, these balls being then fried in oil.

iii. Husked, split, and then boiled with condiments.

iv. Steeped in cold water till they swell, and then eaten either alone or with salt, or else fried in oil or ghí with chillies.

Pigeon-peas are thus used; they are either—

i. Parched in hot sand and eaten dry, or with salt or oil.

ii. Ground into flour and made into cakes, or ground into meal and mixed with water.

iii. Steeped in cold water and rubbed into a paste, small pieces of the fruit of Cucurbita Pepo, salt, carminative seeds and sometimes asafœtida being added. The mass is made into balls, which are dried in the sun and used in curries; this preparation seems peculiar to Behar, Patná and Purniah Districts.

iv. Boiled in about six times their bulk of water until soft, condiments being then added—turmeric, black pepper, capsicum, and sometimes cumin and coriander fruits, with the leaves of Laurus Cassia and salt. A little broken capsicum with mustard oil or ghí is heated in another vessel and browned, the above preparation of dhal being poured into it.

To these recipes may be added a note as to the desirability of washing most kinds of pulse in cold water, and the useful effect (especially in the case of lentils) of a brief soaking in water to which a little carbonate of soda has been added; the alkaline liquor removes a part of the bitter principle present in the pulse, and is then thrown away.

The proportion of pulse to cereal consumed as food differs much in different parts of India. As examples the following Districts may be named where pulse constitutes from $\frac{1}{4}$ to $\frac{1}{24}$ of the food-stuffs consumed, or else is a mere trivial adjunct:

Húglí,—$\frac{1}{4}$ of the rice.

Cáchár,—with fresh vegetables $\frac{1}{4}$ of the food-stuffs.

Bogra, Purí,—$\frac{1}{5}$ of the food-stuffs.

Murshidábád,—$\frac{1}{8}$ of the rice.

Noákhálí, Dinápur, Gaya, Sáran,—$\frac{1}{8}$ of the food-stuffs.

Dinájpur, Maldah,—$\frac{1}{16}$ of the food-stuffs.

Mánbhúm,—$\frac{1}{20}$ of the food-stuffs.

Tirhút,—$\frac{1}{24}$ of the food-stuffs.

Ranjpur, Dacca,—mere adjuncts.

Farídpur,—many kinds of pulse are eaten when fish is not attainable.

Chittagong,—pulse is eaten by well-to-do natives.

The following tabular statement gives the number of acres under pulse in 1882-83 in the five provinces from which the returns are fairly complete:

	Various Pulses.
North-West Provinces	4,350,664
Punjab	3,664,662
Madras	1,955,946
Bombay	1,699,432
Berar	467,465
	12,138,169

The Order Leguminosæ is the second largest Order of flowering plants: it contains between 6,000 and 7,000 species. A conspectus of the tribes and genera of this Order—so far as the plants described in the following pages are concerned—is here given:

Sub-Order I., PAPILIONACEÆ.

Tribes.	Genera.
ii. GENISTEÆ.	Lupinus.
v. GALEGEÆ.	Cyamopsis.
vi. HEDYSAREÆ—	
Sub-Tribe 5. STYLOSANTHEÆ.	Arachis.
vii. VICIEÆ.	Cicer, Vicia, Lathyrus, Pisum, Lens.
viii. PHASEOLEÆ—	
Sub-Tribe 1. GLYCINEÆ.	Glycine.
" 2. ERYTHRINEÆ.	Mucuna.
" 3. GALACTIEÆ.	Canavalia.
" 4. EUPHASEOLEÆ.	Phaseolus, Vigna, Dolichos, Psophocarpus.
" 5. CAJANEÆ.	Cajanus.
Sub-Order II., CÆSALPINIEÆ—	
xiv. CASSIEÆ.	Ceratonia.
xvi. AMHERSTIEÆ.	Tamarindus.
Sub-Order III., MIMOSEÆ—	
xxii. ACACIEÆ.	Pithecolobium.

The sequence in which the several kinds of Indian pulse are arranged and described in the following pages is that indicated in the above table, except that a few species, chiefly belonging to Mucuna, Psophocarpus, and Tamarindus, are relegated to the closing paragraphs of the present part of the work, because no detailed chemical analyses of these plants have been made.

In classifying the several species of pulse according to their chemical composition, the most useful plan to follow is one based on their relative richness in albuminoids, or rather on the value of their nutrient-ratios, that is, the proportion between their albuminoids and their starch, reckoning, however, with their starch, the starch-equivalent of the oil they contain. In the following table the first column of figures gives the above-named nutrient-ratio; the second column the percentage of albuminoids;

and the third the "nutrient-value," or, in other words, the added percentages of albuminoids, of starch, and of oil translated into its starch-equivalent.

	Name of Pulse.	Nutrient-ratio, or Albuminoids to Starch.	Albuminoids, percentage of.	Nutrient-value.
1.	Lupines ...	1 : 1·4	31·7	77
2.	Vetches ...	1 : 1·6	31·5	81
3.	Cyamopsis-beans	1 : 1·7	29·8	79
4	Vetchlings	1 : 1·75	31·9	87
5.	Soy-beans	1 : 2·0	35·3	105
6.	Sword-beans	1 : 2·2	25·0	80
7.	Peas ...	1 : 2·4	23·6	81
8.	Lablab-beans	1 : 2·5	22·4	80
9.	Moth-beans	1 : 2·5	23·8	81
10.	Lentils ...	1 : 2·5	24·9	87
11.	Vigna-beans	1 : 2·5	23·1	81
12.	Haricot-beans	1 : 2·5	23·0	75
13.	Horse-gram	1 : 2·7	22·5	83
14.	Múng-beans	1 : 2·7	22·9	83
15.	Pigeon-peas	1 : 3·0	20·3	80
16.	Lima-beans	1 : 3·2	19·8	80
17.	Chick-peas	1 : 3·3	19·5	84
18.	Inga-beans	1 : 4·6	17·6	98
19.	Pea-nuts...	1 : 5·2	24·5	151
20.	Carob-beans	1 : 8·5	7·1	68

THE LUPINE.

Lupinus albus, L.

Hind.—Turmas. *Punjab*—Bakla-misrie.

This well-known annual herb, a native of the Levant, is cultivated in some parts of India. It is much grown in some parts of Europe, particularly in Germany, where several varieties of L. albus, and two or three other species, are cultivated mainly for the sake of the seeds.

COMPOSITION OF LUPINES (C.).

	In 100 parts.	In 1 lb.
Water	12·5	2 oz. 0 grs.
Albuminoids	31·7	5 „ 32 „
Starch	33·7	5 „ 171 „
Oil...	5·0	0 „ 350 „
Fibre	13·5	2 „ 70 „
Ash	3·6	0 „ 252 „

The nutrient-ratio in lupines is 1 : 1·4, and the nutrient-value 77. The high amount of fibre present in these seeds renders a good deal of the albuminoids and oil they contain unavailable for digestion. It should also be noted that the seeds of most varieties of lupine contain one or more bitter principles. These contain nitrogen, are of the nature of alkaloids, and are poisonous. The seeds of the blue lupine contain a liquid alkaloid, different from the two alkaloids found in the yellow lupine. The seeds of some few varieties may perhaps be free from these objectionable constituents, but careful and complete experiments to determine this point are still wanting. The poisonous alkaloids of lupines are dissipated or destroyed when the seeds are thoroughly cooked by boiling.

GUAR-BEANS.

Cyamopsis psoralioides, D.C.

Synonyms—Psoralea tetragonoloba (Linn.); Lupinus trifoliatus (Cav.); Dolichos psoralioides (Lamk.); D. fabæformis (L'Hérit.).

Hind.—Mutkí-pulli. *Punjab*—Mot-safaid, Guár. *Tam.*—Koth-averay. *Tel.*—Goruchik-kudu.

A robust, hairy, erect annual herb, 2 to 3 feet high. It belongs to the sub-order Papilionaceæ of the Leguminosæ and to the tribe Galegeæ. The other species of Cyamopsis is African. The pods of this species are thick, and from 1½ to 2 inches long. It grows on the plains from the Himálaya to the Western Peninsula. It is cultivated, not only for its ripe seeds, but as a green vegetable, the pods being eaten like those of French beans. The dry beans are used as food for animals as well as for man; an average yield is 10 maunds per acre.

COMPOSITION OF GUAR-BEANS.

	In 100 parts.	In 1 lb.
Water	11·8	1 oz. 388 grs.
Albuminoids	29·8	4 „ 336 „
Starch	46·2	7 „ 171 „
Oil	1·4	0 „ 98 „
Fibre	7·7	1 „ 102 „
Ash	3·1	0 „ 217 „

The nutrient-ratio is here 1 : 1·7, while the nutrient-value is 79; the indigestible fibre is rather high.

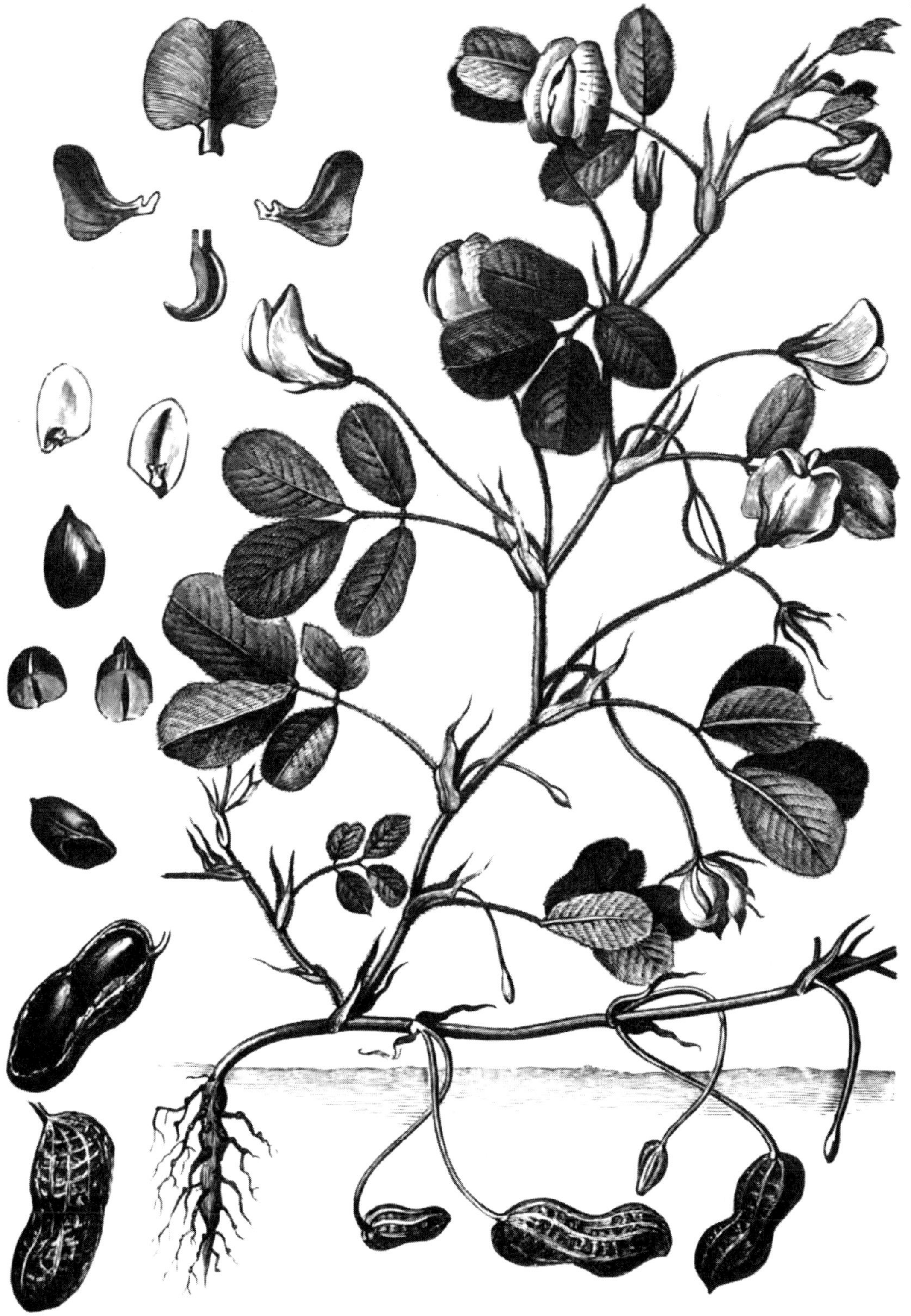

FIG. 22. PEA-NUT (*Arachis hypogæa*).

THE PEA-NUT.

Arachis hypogæa, L.

Hind.—Múng-phullie, Búe-múng, Vilayeti-múng. *Punjab*—Chawal múgra. *Beng.* —China-badano, Alke-kulay. *Tel.*—Nela-sanagalu. *Tamil*—Nelay-cadalay. *Sanskrit*—Bochanaka.

The ground or pea-nut is an annual herb with procumbent branches. It belongs to the sub-order Papilionaceæ, the tribe Hedysareæ, and the sub-tribe Stylosantheæ. It presents somewhat the appearance of a large kind of clover; it has small, bright yellow, pea-like flowers borne on long stalks; these, after flowering, curl downwards and force the immature pod into the soil, where it ripens. The pod is about 1½ inch long and contains from 2 to 3 seeds, 75 of which weigh one ounce. This plant is probably of American origin, although it has been long cultivated in India, on the West Coast of Africa, and in many other tropical countries. There is a similar plant, Voandzeia subterranea, allied to Vigna, which grows under the same conditions.

COMPOSITION OF PEA-NUTS (C.).

	In 100 parts.	In 1 lb.
Water	7·5 ...	1 oz. 87 grs.
Albuminoids	24·5 ...	3 „ 403 „
Starch	11·7 ...	1 „ 382 „
Oil	50·0 ...	8 „ 0 „
Fibre	4·5 ...	0 „ 315 „
Ash	1·8 ...	0 „ 126 „

The nutrient-ratio is here 1 : 5·2, and the nutrient-value so high as 151. As half the weight of pea-nuts is oil, they require a considerable admixture of starchy food in order to become a wholesome and economical article of diet. The green and unripe pods are less oily and more easily digested; they have an agreeable taste when roasted. Pea-nuts, after the greater part of the oil has been extracted by pressure, yield a cake well adapted for feeding cattle.

THE CHICK-PEA.

Cicer arietinum, L.

Hind.—Chola, Channa, Chana, Hurbari, Adas, Chela, Rohala. *Beng.*—Bút, Chala, Chuna. *Tam.*—Cadalei. *Telugu*—Harimandha-kam. *Sanskrit*—Chennuka.

A viscose much-branched annual, with pinnate leaves usually having a terminal leaflet. The plant attains a height of 2 feet or more. The pods are ¾ to 1 inch long, and usually contain two seeds which are generally somewhat symmetrically crinkled: about 90, sometimes less, weigh one ounce. Of the 7 species of Cicer 2 are Indian. The genus Cicer belongs to the tribe Vicieæ, and the sub-order Papilionaceæ.

The chick-pea or common gram is largely cultivated in the Northern Provinces and the Nilgiris. It is sown in August or September, and reaped in April and May; if sown in October or November, it is ready in February, March, or April. The plants should be 1 foot apart, 2 feet being left between the rows. The ripe, unhusked seeds are largely used for horses and cattle in many parts of India; the seeds, after parching, grinding, steeping, or the removal of the husks, form an important food in some districts, chiefly for the labouring class. The young leaves fried in oil or ghí are also considered wholesome: they are sometimes ordered as the exclusive diet in leprosy by native physicians.

The figures given below are the means of 9 analyses of the unhusked peas and of 4 analyses of the peas from which the husk has been removed.

COMPOSITION OF THE CHICK-PEA.

	In 100 parts. Husked.	In 100 parts. With Husk.	In 1 lb. Husked.
Water	11·5	11·2	1 oz. 367 grs.
Albuminoids	21·7	19·5	3 „ 207 „
Starch	59·0	53·8	9 „ 192 „
Oil	4·2	4·6	0 „ 294 „
Fibre	1·0	7·8	0 „ 70 „
Ash	2·6*	3·1†	0 „ 182 „

* 1·1 of phosphoric acid. † 0·8 of phosphoric acid.

FIG. 23. CHICK-PEA (*Cicer arietinum*).

The nutrient-ratio in the unhusked peas is 1 : 3·3 ; the nutrient-value is 84.

When growing, this pea exudes a sour secretion which contains acid oxalate of potash, and other acid salts.

The second Indian species of Cicer is C. soongaricum, the leaves of which are 3 to 4 inches long, the leaflets in 10 to 12 pairs, the end a spiral tendril. It belongs to the temperate and alpine region of the Western Himálaya, from 9,000 to 15,000 feet: Piti Lahúl, Kumáun, Tibet, etc. I have not succeeded as yet in obtaining a sample of this species of chick-pea for analysis.

THE VETCH.

Vicia sativa, L.

Hind.—Anhuri. *Lahore*—Mattz-rewari.

A slender annual belonging to the tribe Vicieæ. Its leaves have 8 to 12 leaflets; the pod is 1½ to 2 inches in length, 8 to 10-seeded. It occurs in the North-West Provinces, from the plains of Bengal up to 7,500 feet in Kumáun, and is probably not anywhere truly wild but always cultivated, though it is said to occur spontaneously in the Patná District, where, though little used or esteemed, it is made into cakes by the poor. V. narbonensis occurs in the Punjab near Pesháwar. It is probably an introduction.

COMPOSITION OF VETCHES.

	In 100 parts.	In 1 lb.
Water	10·1	... 1 oz. 270 grs.
Albuminoids	31·5	... 5 " 17 "
Starch	47·6	... 7 " 270 "
Oil	·9	... 0 " 63 "
Fibre	6·7	... 1 " 31 "
Ash	3·2	... 0 " 224 "

The albuminoids in the above analysis of Indian vetches are rather high, the average of European samples giving 27·5 per cent. only. The nutrient-ratio in the Indian vetches is 1 : 1·6.

The garden and field beans of Europe are also grown in India; they are the seeds of Vicia Faba (L.), commonly known as Faba vulgaris. They contain on an average about 25 per cent. of albuminoids and 7·5 per cent. of fibre. These beans are grown to a considerable extent in the North-West Provinces.

THE VETCHLING.

Lathyrus sativus, L.

Synonym—Cicercula alata (Moench.).
Hind.—Khesári, Kussúr, Kasári, Kassar-tiuri, Latri. *Beng.*—Teyuri, Teora. *Punjab*—Churál. *Sind*—Matar.

A much-branched annual herb, having equally pinnate leaves; leaflets 2, linear or lanceolate. The pods are 1½ inch long, 4 to 5-seeded. It is spread through the Northern Provinces, ascending from the plains of Bengal to 4,000 feet in Kumáun.

The genus Lathyrus belongs to the tribe Vicieæ of the sub-order Papilionaceæ. There is another species, not an Indian plant (L. tingitanus), which like L. sativus is extensively cultivated.

This is a cold-weather or rabi crop, and is grown on land unfitted for most other pulse. It is sown in October and November and reaped in March and April.

COMPOSITION OF VETCHLINGS.

	In 100 parts.	In 1 lb.
Water	10·1 ...	1 oz 270 grs.
Albuminoids	31·9 ...	5 „ 45 „
Starch and Fibre	53·9 ...	8 „ 273 „
Oil	·9 ...	0 „ 63 „
Ash	3·2 ...	0 „ 224 „

The nutrient-ratio is here about 1 : 1·75, while the nutrient-value is nearly 87. There is reason to suspect the occasional presence, in injurious proportion, of a poisonous bitter principle in this vetchling. It has a bad reputation, and is almost universally regarded in Bengal as unwholesome, deranging digestion, and producing dysentery, diarrhœa, and various skin diseases. But some allowance must be made for the prejudice of the Bengalese. It is most used by the poorer classes, being

133 FIG 24. VETCHLING (*Lathyrus sativus*).

the cheapest and most abundant pulse. Many cases of sudden and incurable paralysis have been undoubtedly traced to the large and continuous use of this seed. It formed, by a series of accidents, the chief food, during the years 1829–33, of some of the eastern villages of Oudh. Many cases of sudden paralysis of the lower extremities occurred during that period, the persons attacked being generally under thirty years of age.

This is a coarse kind of pulse, hard and difficult to cook. It is used in Behar and Patná in curries. It is also made into paste-balls which are fried in ghí and eaten with boiled rice; it is also eaten as dal.

THE PEA.

Pisum sativum, L.

Hind.—Mattar, Gol-mattar, Buttaní-chola. *Beng.*—Bura-mattar, Kuda. *Tamil*—Pattanie.
Sanskrit—Harenso.

The garden pea, a familiar annual herb, the leaves having three pairs of entire glaucous leaflets, and large leafy stipules. Pisum belongs to the tribe Vicieæ. As a wild plant P. sativum is regarded as a native of the South Caucasus to Persia. It has been long cultivated in India.

The pea prefers heavy ground, and generally receives little cultivation, manure or irrigation. The produce of ripe seeds varies from 7 to 8 maunds per acre, without irrigation; on irrigated land the yield is increased to 10 up to 16 maunds. The green pods are largely eaten before the general crop is cut in February, March, or April; the sowing takes place in October and November.

COMPOSITION OF PEAS.

	In 100 parts. Husked.	In 100 parts. Unhusked.	In 1 lb.	
Water	11·8	12·5	2 oz.	0 grs.
Albuminoids	28·2	23·6	3 „	340 „
Starch	55·0	54·5	8 „	315 „
Oil	1·5	1·3	0 „	91 „
Fibre	1·0	5·7	0 „	399 „
Ash	2·5*	2·4†	0 „	168 „

* 1·0 of phosphoric acid. † 0·8 of phosphoric acid.

The mean of a large number of analyses, made of peas grown in Europe, shows an almost complete accordance with the above results, which yield a nutrient-ratio of 1 : 2·4 and a nutrient-value of 81, in the unhusked pulse.

In some districts this pea is not much esteemed in comparison with other pulse. Undoubtedly it has a tendency to produce flatulence, and is unwholesome when imperfectly cooked or imperfectly freed from the tough coat of the seed. In the south of the Purniah District it is largely eaten uncooked, and causes or aggravates dysentery and diarrhœa.

The field pea, sometimes called Pisum arvense, is a variety or sub-species of this species. It is sown and reaped at the same time as the garden pea. It is rightly regarded as more indigestible than the latter, but there is no constant difference of chemical composition between the two sorts. The garden pea has round seeds and 4 to 6 leaflets, the field pea marbled compressed seeds and 2 to 4 leaflets.

For methods of preparing these peas for food see page 120.

THE LENTIL.

Lens esculenta, Moench.

Synonyms—Ervum lens (L.); Cicer lens (Willd.).
Hind.—Masúri, Mussúr. *Beng.*—Buro-Mussúr, Mussúri.

This plant belongs to the tribe Vicieæ; the botanical name by which it is best known is Ervum lens, but the genus Ervum has now been sunk, partly in Lens. The lentil is a branched annual with oblong leaflets, usually 8 in number. The pod is broad and short, and contains 2 seeds, weighing from 1 to 1½ grain apiece in the large seeded variety. The seeds are compressed, and have the form of a bi-convex lens. This plant has been largely cultivated from very ancient times; its native country is unknown.

The lentil may be grown on almost all soils; it flourishes upon those which, while light, lie low. It is grown like peas as

137 Fig. 25. Lentil (*Lens esculenta*).

a cold-weather crop, being sown in September and October, and reaped in March and April. It is commonly cultivated, especially in the North-West Provinces and Madras. It yields from 6½ to 8 maunds per acre, or, if irrigated, 10 to 12 maunds. The yield might be increased if more pains were taken in the selection of seed for sowing, as there are some varieties of the lentil which produce seeds weighing twice as much as the small common sort, and which yet do not make a proportionately increased demand upon the resources of the soil.

COMPOSITION OF LENTILS.

	In 100 parts. Husked.	In 100 parts. With Husk.	In 1 lb.
Water	11·8	11·7	1 oz. 380 grs.
Albuminoids	25·1	24·9	3 „ 430 „
Starch	58·4	56·0	8 „ 420 „
Oil	1·3	1·5	0 „ 105 „
Fibre	1·2	3·6	0 „ 252 „
Ash	2·2*	2·3†	0 „ 161 „

The nutrient-ratio in lentils is 1 : 2·5, and the nutrient-value 87. The lentil is generally regarded as a pulse of the second class, inferior to múng (*Phaseolus Mungo*), but equal to urhur, the pigeon-pea. It is highly nutritious but somewhat heating; it should be carefully freed from the husk or coat. The bitter substance which occurs in lentils may be removed to some extent by soaking them for a short time in water in which a little carbonate of soda (common washing soda) has been dissolved. The meal of lentils, deprived of their coat, is of great richness, containing generally more albuminoid or flesh-forming matter than bean or pea-flour. The preparations advertised under the names of "Revalenta," "Ervalenta," etc., consist mainly of lentil meal, mixed with the flour of barley or some other cereal, and common salt.

* 0·8 of phosphoric acid. † 0·7 of phosphoric acid.

THE SOY-BEAN.

Glycine Soja, Sieb. and Zucc.

Synonyms—Soja hispida (Moench.); Dolichos Soja (Linn.); Soja angustifolia (Miq.). *Hind.*—Bhat, Bhatwan. *Punjab*—Bhút. *Beng.*—Gari-kulay. *Naga*—Tsu-dza.

This important bean is the seed of Glycine Soja, a small, sub-erect, trifoliate, hairy annual, with pods generally 3 to 4-seeded. It belongs to the natural order Leguminosæ, sub-order Papilionaceæ, tribe Phaseoleæ, and sub-tribe Glycineæ: 5 genera are included in this sub-tribe. Glycine contains about 12 species, chiefly Australian, but 3 are Indian, namely, G. javanica, G. pentaphylla, and our present species.

The soy-bean forms a considerable article of food in China and Japan. Since 1873 it has been successfully grown, as an experiment, in some of the warmer parts of Europe. It is widely spread in the outer Himálaya, and tropical regions from Kumáun to Sikkim, and the Khásir, and the Naga Hills to Upper Burma. It is often cultivated, rather largely in Bustí and Gorakhpur, Patná and Purniah Districts.

This crop is generally grown by itself; the seeds are sown from June to September; the harvesting takes place between November and January. It is consequently a kharif crop. The seeds should be placed at a depth not exceeding 1 to 1½ inch; 18 plants may be left, after weeding and thinning, to the square yard. A peaty soil, or one rich in organic matter, suits the plant best; a calcareous soil is also favourable to its growth. Sulphate of potash is a good manure; nitrogen may be supplied either as nitrate of soda, or, in the case of soils poor in organic matter, in the form of rape or mustard cake, but it is rarely needed, while large applications of nitrogenous manure exert a distinctly injurious effect upon the yield of beans. So far as we know, this very important, vigorous, and productive pulse is not attacked by any insect or parasitic fungus.

Although there are a number of varieties of the soy-bean, the chief differences between them lying in the size, shape, and colour

141 FIG. 26. SOY-BEAN (*Glycine Soja*).

of the seeds, yet these varieties are not distinguished by definite differences in chemical composition. That composition entitles the soy-bean to the highest place, even amongst the pulses, as a food capable of supplementing the deficiencies of rice and of other eminently starchy grains. Very few vegetable products are so rich as this bean at once in albuminoids and in fat or oil, the former constituent amounting on the average to 35 per cent., and the latter to 19. The cultivation of the pale large-seeded varieties should be extended.

COMPOSITION OF SOY-BEANS (C.).*

	In 100 parts.	In 1 lb.
Water	11·0	1 oz. 333 grs.
Albuminoids	35·3	5 ,, 283 ,,
Starch and Sugar	26·0	4 ,, 70 ,,
Fat	18·9	3 ,, 10 ,,
Fibre	4·2	0 ,, 294 ,,
Ash	4·6	0 ,, 322 ,,

The nutrient-ratio is here about 1 : 2, while the nutrient-value is 105. Potash forms nearly one-half, and phosphorus-pentoxide one-third of the ash of the soy-bean. Ripe soy-beans require long soaking, preferably in warm water, in order to render them soft.

In China and Japan three preparations are extensively made from the soy-bean. Soy sauce is the best known of these, but more important are the soy or bean cheeses, and a kind of paste. The beans are sometimes pressed for the sake of the oil they yield ; the residual cake forms an extremely rich cattle food, con-

* The mean percentages, deduced from 8 analyses of unhusked soy-beans, 4 of the samples being of Chinese origin, and from 2 of husked soy-beans, are thus given by Dr. Forbes Watson :

	With Husk.	Husked.
Water	9·1	10·3
Albuminoids	40·4	43·6
Starch and Sugar	25·1	21·0
Fat	15·8	15·5
Fibre	5·2	4·4
Ash	4·4	5·2

taining as it does 40 per cent. of flesh-formers and 7 per cent. of oil. The soy-bean may also be grown as a fodder plant. If cut just when the pods are fully formed it makes an excellent hay, superior to that of the lentil.

THE SWORD-BEAN.

Canavalia ensiformis, D.C.

Synonyms—Canavalia gladiata (D.C.); C. incurva (D.C.); Dolichos ensiformis (L.); D. gladiatus (Jacq.); C. virosa (W. & A.); C. Stocksii (Dalz.); C. mollis (W. & A.).

Hind.—Kudsumber, Mukhum-síno, Chotu-síno. *Beng.*—Mukshimo-shíno. *Tamil*—Segapú, Velay-thumbettan, Coli-averakai. *Telugu*—Chamma, Yerra-tumbettan-kaya. *Deccan*—Chotie-saymke-pullie. *Sinhalese*—Awara. *Mysore*—Kakara-parang. *Burma*—Pai-noung-neo.

A twining perennial or biennial herb with trifoliate leaves, the leaflets being ovate or oblong, and from 4 to 6 inches in length. The pod is 8 to 12-seeded, 6 to 9 inches long (and even more), and 1 to 1¼ inch broad.

This species of Canavalia extends from the Eastern Himálaya to Ceylon and Siam; it is commonly cultivated.

COMPOSITION OF SWORD-BEANS (C.).

	In 100 parts.	In 1 lb.	
Water	12·5 ...	2 oz.	0 grs.
Albuminoids	25·0 ...	4 „	0 „
Starch	48·6 ...	7 „	339 „
Oil	2·8 ...	0 „	196 „
Fibre	7·7 ...	1 „	102 „
Ash	3·4 ...	0 „	238 „

The nutrient-ratio is here 1 : 2·2, and the nutrient-value 80.

 FIG. 27. SWORD-BEAN (*Canavalia ensiformis*).

HARICOT-BEANS.

Phaseolus vulgaris, L.

Hind.—Bakla, Loba.

A well-known annual, having sub-erect or twining stems 6 to 10 feet long, and pods 4 to 6 inches long, with 4 to 6 seeds. Many cultivated forms of this plant have been described.

The ripe seeds of many varieties of this species are known in Europe as haricot-beans. The unripe pods are largely used as a green vegetable under the names of French beans and kidney-beans. Amongst the very numerous varieties of haricots cultivated in Europe as a green vegetable, one of the best is the Haricot Beurre of the French. The pods of the scarlet-runner (*Ph. multiflorus*), or Haricots d'Espagne, are similar, but the ripe seeds are often unwholesome.

COMPOSITION OF HARICOT-BEANS.

	In 100 parts.	In 1 lb.
Water	14·0 ...	2 oz. 105 grs.
Albuminoids	23·0 ...	3 „ 297 „
Starch	52·3 ...	8 „ 161 „
Oil...	2·3 ...	0 „ 161 „
Fibre	5·5 ...	0 „ 385 „
Ash	2·9 ...	0 „ 203 „

The nutrient-ratio is 1 : 2·5, and the nutrient-value 75.

Besides the species of Phaseolus here described or mentioned three others are cultivated in India, namely: Ph. calcaratus, Ph. semi-erectus, and Ph. lunatus. The first of these is commonly cultivated in the tropical zone; it has a narrow, recurved, 8 to 12-seeded pod. Ph. semi-erectus has a longer and still narrower pod with many seeds; it is spread through the Western Peninsula and Ceylon. The third of these species, Ph. lunatus, is a very different kind of bean; a brief account of it will be found on page 155.

THE MÚNG-BEAN (includes BLACK GRAM and GREEN GRAM).

Phaseolus Mungo.

Synonyms—Phaseolus Mungo (Roxb.); P. Max (Roxb.); P. aureus (Ham.).
Hind.—Dord, Múng, Thikirí, Múg. *Beng.*—Mash-kulay, Múg. *Punjab*—Másh, Urd. *Madras*—Pessalú.
Sanskrit—Múdga, Masha.

This species of Phaseolus includes a number of forms to several of which specific rank has been accorded by some botanists; a typical form and 3 varieties are now recognised. The type includes P. Max (Roxb.), with black seeds, P. aureus (Ham.), with yellow seeds, and the green-seeded P. Mungo of Roxburgh. The typical form has sub-erect or flexuose stems, which with the pods are densely hairy. The varieties are:

Var. 1, glaber (Roxb.), having the habit of the type but with glabrous stems, leaves, and pods. It is the P. glabrescens of Steudel.

Var. 2, wightianus (Grah.), having elongated slender stems. It is P. Wightii (W. & A.), and P. subvolubilis (Ham.).

Var. 3, radiatus (L.), having elongated twining stems densely clothed. It is identical with P. Roxburghii (W. & A.), and P. setulosus (Dalz.), and near the species P. trinervius (Heyne), which may after all be a fourth form of Phaseolus Mungo.

The pod of this species is 1½ to 2½ inches long, by ⅛ to ⅕ inch broad; it contains 10 to 15 seeds, and is slightly recurved. The seeds vary much in colour and a good deal in size; they are sometimes dull, sometimes shining.

The type form is wild, and universally cultivated, ascending to 6,000 feet in the North-West Himálaya. The var. radiatus exists in two forms: one, having large blackish seeds, ripens (according to Mr. Duthie) in the North-West Provinces and Oudh, in September; the small green-seeded sort is gathered in October and November. According to the same authority this variety prefers stiff soils, while the typical form is generally grown on light sands. The yield is about 5 maunds per acre.

149 FIG. 28. MUNG-BEAN (*Phaseolus Mungo*).

Of this bean in its many varieties, a large number of analyses have been made without disclosing any decided differences in chemical composition. However, other things being equal, preference should be given to the larger-seeded varieties. This plant withstands drought well, and forms a valuable food resource when millets fail. It is ruined by heavy rains during its flowering. It is sown, according to locality, in June to September, and reaped from September to December.

COMPOSITION OF MÚNG-BEANS (C.).

	In 100 parts, with Husk.			
	Type—Green Seeds.	Type—Yellow Seeds.	Var.—Radiatus.	In 1 lb.
Water	10·8	11·4	10·1	1 oz. 270 grs.
Albuminoids	22·2	23·8	22·7	3 „ 276 „
Starch	54·1	54·8	55·8	8 „ 406 „
Oil	2·7	2·0	2·2	0 „ 154 „
Fibre	5·8	4·2	4·8	0 „ 336 „
Ash	4·4	3·8	4·4*	0 „ 308 „

The nutrient-ratio of the unhusked beans is 1 : 2·7, the nutrient-value 83. The fibre in the husked beans is reduced to 1·1 per cent., all the other constituents being proportionately increased.

In most localities this bean is esteemed highly, and is regarded as wholesome; it is generally eaten by the richer classes, and is resorted to by all, wherever possible, in times of sickness. By some authorities it is stated to cause flatulence unless eaten with asafœtida. But it must be remembered that all pulse has this tendency, if it form too large a proportion of the day's ration or be insufficiently cooked.

It is of interest to note that the ash of the straw or stems and leaves of one of the varieties of this bean (var. 3, radiatus) is sometimes eaten, in Dinajpur for example, in lieu of salt. The craving for mineral matter brought about by the marked deficiency of rice in ash-constituents, is partially satisfied by the use of the ashes not only of this pulse, but of several other kinds.

---

* Includes 1·1 of phosphoric acid.

THE MOTH-BEAN.

Phaseolus aconitifolius. Jacq.

Synonyms—Ph. trilobus (Wall.), Dolichos dissectus (Lam.).
Hind.—Mút, Mote, Moth, Mothi, Bhringga, Meth-kalai. *Beng*—Kheri. *Assam*—Matti-kalaie. *Tamil*—Tulka-pyre. *Telugu*—Kúncúma-pesalú. *Sind*—Mohar.

This bean is the produce of a trailing slender-stemmed herb belonging to the tribe Phaseoleæ. It is found from the Himálaya to Ceylon, in the tropical region, and extends up to 4,000 feet in the North-West. It is cultivated in Oudh, Allahábád, Farruckábád, Patná District, Purniah District, Assam, Gorakhpur. The pods are stouter and the seeds larger than in the allied species, Ph. trilobus (Ait.). It is often grown on the worst land which can be made to yield a crop at all. It is frequently sown with bájra, the bulrush millet, on light sandy soils. An average produce is 8 maunds from an acre. It is not esteemed as a food for man, for although it is rich in nutrients, it is generally thought to possess heating properties. It is a kharif crop, being sown in June, July, August, or September, and reaped in November, December, and January.

COMPOSITION OF MOTH-BEANS.

	In 100 parts.	In 1 lb.
Water	11·2 ...	1 oz. 347 grs.
Albuminoids	23·8 ...	3 „ 353 „
Starch	56·6 ...	9 „ 25 „
Fat...	·6 ...	0 „ 42 „
Fibre	4·2 ...	0 „ 294 „
Ash	3·6* ...	0 „ 252 „

The nutrient-ratio is here about 1 : 2·5, while the nutrient-value is 81.

* 0·8 of phosphoric acid.

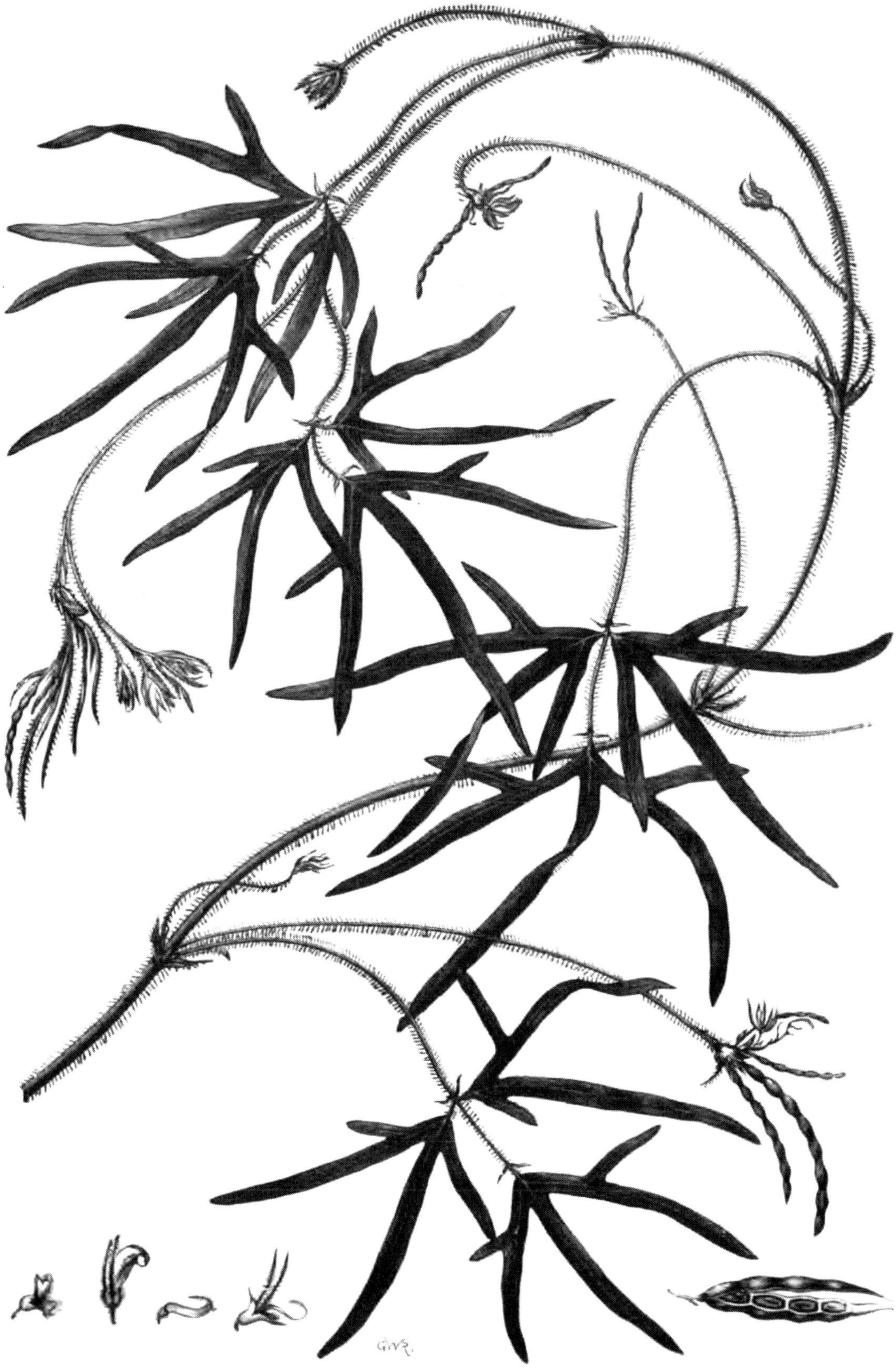

FIG. 29. MOTH-BEAN (*Phaseolus aconitifolius*).

THE LIMA OR DUFFIN-BEAN.

Phaseolus lunatus, L.

Hind.—Kursumbulle-pullie, Bunbur-buttí. *Punjab*—Lobiya.

A tall twining biennial herb, easily distinguished by the pods from other species of Phaseolus. The pod is 2 to 3 inches long, by 3/8 to 5/8 inch broad The seeds are large, but variable both in colour, markings, and size. This plant is cultivated almost everywhere throughout India, but is probably a native of America.

A white variety of seed (from Mysore) was selected for analysis. These seeds averaged 16½ grains in weight apiece. The seeds of a common black-veined variety weigh about 7 grains each.

COMPOSITION OF LIMA OR DUFFIN-BEANS (C.).

	In 100 parts.	In 1 lb.
Water	13·3 ...	2 oz. 56 grs.
Albuminoids	19·7 ...	3 „ 67 „
Starch	57·8 ...	9 „ 108 „
Oil...	1·2 ...	0 „ 84 „
Fibre	4·3 ...	0 „ 301 „
Ash	3·7 ...	0 „ 259 „

The nutrient-ratio is here 1 : 3·2, and the nutrient-value 80. This is one of the species of Phaseolus which sometimes exhibits marked poisonous properties.

It is desirable that great care should be taken in selecting for cultivation the best variety of Lima-beans. The large oval white-seeded kinds, with at the most a brown or black mark close to the hilum, are preferable to those with flattened, rather reniform seeds having blotches of red or veinings of black.

CATIANG-BEANS.

Vigna Catiang (Endl.).

Synonyms—Dolichos catiang (L.); D. sinensis (L.); D. melanophthalmus (D.C.).
Hind.—Lobia, Rawás, Rausa, Souta, Bora. *Beng.*—Barbati. *Deccan.*—Chowlí.
Punjab—Rawan, Souta.
Sanskrit—Lasunda, Raja-masha.

Vigna is a genus of Euphaseoleæ, a sub-tribe of Phaseoleæ. V. Catiang includes many cultivated varieties, differing much in habit of growth as well as in the shape, size, and colour of the seed. The plant is sometimes low and sub-erect (typical V. Catiang), sometimes (var. sinensis) tall and voluble. The pods in the cultivated forms are under ½ inch broad, but may attain 1 foot up to 2 feet in length; the pods contain from 10 to 20 seeds. It is a native plant, and is universally cultivated in the tropical zone. It flourishes in comparatively poor soils and sustains fairly well a considerable degree of drought. Vigna Catiang is sown in July and August, and reaped in October and November. The white-seeded sort is generally considered the best. It is often grown with other crops. An ounce weight corresponds to about 150 seeds.

COMPOSITION OF CATIANG-BEANS.

	In 100 parts.		In 1 lb.
	Husked.	With Husk.	
Water	12·5	12·7	2 oz. 14 grs.
Albuminoids	24·1	23·1	3 „ 305 „
Starch	56·8	55·3	8 „ 371 „
Oil	1·3	1·1	0 „ 77 „
Fibre	1·8	4·2	0 „ 294 „
Ash	3·5*	3·6†	0 „ 252 „

The nutrient-ratio in these (unhusked) beans is therefore 1 : 2·5, and the nutrient-value 81. They are thought to be rather heating and less digestible than úrd or múng (Phaseolus Mungo). The green pods are sometimes cooked.

* 1·0 of phosphoric acid. † 1·2 of phosphoric acid.

157 FIG. 30. VIGNA-BEAN (*Vigna Catiang*).

FIG. 31. LABLAB-BEAN (*Dolichos Lablab*).

THE LABLAB-BEAN.

Dolichos Lablab, L.

Synonym—Lablab vulgaris (Savi).
Hind.—Sim, Pertab-sing. Sém, Sémbi, Shimbi. *Beng.*—Bunsim, Gúrdal-shím. *Telugu*—Annapa, *Tamil*—Mutcheh.
Sanskrit—Nespava, Shimbi.

This wide-twining perennial, or, in cultivation, annual herb, belongs to the same sub-tribe of the Phaseoleæ as Vigna and Phaseolus. It is a true native of India, ascending to 6,000 feet in the Himálaya, and is in universal cultivation. The stems sometimes attain a length of 16 to 20 feet; the recurved flat oblong pod is 1½ to 2 inches long by ½ to ¾ inch broad. It is 2 to 4-seeded; there are about 150 seeds to an ounce. Many varieties have been fully described and have received specific names.

This pulse is grown not only for its ripe seeds but also for its green pods, which are used as a vegetable. In some places it is grown with castor-oil plants, as if alone it would need stakes as a support.

COMPOSITION OF LABLAB-BEANS.

	In 100 parts. With Husk (1, C.).	With Husk (2, C.).	Husked (3).	With Husk (4).	In 1 lb.
Water	14·6	14·3	12·1	12·1	1 oz. 410 grs.
Albuminoids	17·1	20·5	24·4	22·4	3 „ 255 „
Starch	57·4	53·5	57·8	54·2	8 „ 294 „
Oil	2·3	2·2	1·5	1·4	0 „ 98 „
Fibre	5·0	5·8	1·2	6·5	1 „ 17 „
Ash	3·6	3·7	3·0	3·4*	0 „ 238 „

The nutrient-ratio, deduced from analysis 4, is 1 : 2·5, the nutrient-value is 80. It will be seen, however, on comparing the several analyses given above, that the percentage of albuminoids is rather variable. The extreme range is probably not more than 6 per cent. Of the numerous forms of Lablab the majority are eaten as a green vegetable.

* 1·0 of phosphoric acid.

HORSE GRAM.

Dolichos biflorus, L.

Synonym—Dolichos uniflorus (L.).
Hind.—Kúlthí, kúltí. *Punjab*—Kalat, Gagli, Bárat, Botang. *Tamil*—Wúla-walú.
Sanskrit—Kolutha.

This species of Dolichos is either sub-erect or twining in habit. The pod is much recurved, 1½ to 2 inches in length, 5 to 6-seeded. It is wild in the Himálaya to Ceylon and Burma, ascending to 3,000 feet in Sikkim. It is not infrequently cultivated, being generally sown from August to November, and reaped from November to February.

COMPOSITION OF HORSE GRAM (C.).

	In 100 parts. Unhusked.	In 1 lb.
Water	11·0	1 oz. 333 grs.
Albuminoids	22·5	3 „ 262 „
Starch	56·0	8 „ 420 „
Oil	1·9	0 „ 133 „
Fibre	5·4	0 „ 378 „
Ash	3·2	0 „ 224 „

The nutrient-ratio is here 1 : 2·7, and the nutrient-coefficient 83. The ash of these beans contains nearly one-third of its weight of phosphoric acid. The long-continued use of these beans is regarded as injurious; they are reputed, in some districts, to cause œdematous swellings. The haulms are a good fodder.

FIG. 32. HORSE GRAM (*Dolichos biflorus*).

FIG. 33. HORSE GRAM (*Dolichos biflorus*).

FIG. 34. PIGEON-PEA (*Cajanus indicus*).

THE PIGEON-PEA.

Cajanus indicus, Spreng.

Synonyms—Cytisus Cajan (L.); Cajanus flavus (D.C.); Cajanus bicolor (Wall.).
Hind.—Arhar, Thúr, Dal, Burrie-túvar, Arhuku, Shakul. *Beng.*—Dal-urur, Orol.
Tamil—Thovaroy. *Telugu*—Kandalú.
Sanskrit—Adaki, Arhuka.

The pigeon-pea belongs to the sub-tribe of the Cajaneæ, the 5th under the tribe Phaseoleæ. There is but one species of Cajanus, and that is not truly Indian but African, although it has been long cultivated in India, as the existence of a Sanskrit name (arhuka) for it testifies. It is an erect shrub with slender branchlets. The pod is straight, from 2 to 3 inches long, and ¼ to ½ inch broad, 3 to 5-seeded. A variety, C. bicolor, has the yellow standard of the corolla beautifully veined with red.

This plant is extensively grown almost throughout India. It is sown in June or July, and reaped according to locality from December to March. It is commonly grown with juár, bájra, or cotton. It prefers a light but moist soil; it is rarely irrigated. The yield of seeds varies from 7 to 16 maunds per acre when the crop is one of this pea alone.

COMPOSITION OF THE PIGEON-PEA.

	In 100 parts. Husked (1, W.).	Unhusked (2, C.).	Unhusked (3, W.).	In 1 lb.
Water	10·5	13·3	11·4	1 oz. 361 grs.
Albuminoids	22·3	17·1	20·3	3 „ 108 „
Starch	60·9	55·7	56·4	9 „ 11 „
Fat	2·1	2·6	1·4	0 „ 98 „
Fibre	1·2	7·5	7·1	1 „ 59 „
Ash	3·0*	3·8	3·4†	0 „ 238 „

The nutrient-ratio is, in analysis 3 (the mean of 3), about 1 : 3; the nutrient-value is 80.

This pea is largely consumed by all classes in many parts

* 0·8 of phosphoric acid. † 0·9 of phosphoric acid.

of India. In Sáran it is eaten to a larger extent than any other pulse, and is prized next to Moth (*Phaseolus aconitifolius*). It is wholesome and nutritious when properly freed from the husk, its irritant and laxative character being thus greatly reduced. It is not unusual to find that the higher-priced and finer qualities of this pea have been slightly oiled before sale, to improve their appearance. This practice is not unknown in reference to wheat in the South of Europe.

For the modes of preparing the pigeon-pea for human food see the general observations on Pulses, page 120.

THE LOCUST OR CAROB-BEAN.

Ceratonia Siliqua, L.

Punjab—Kharnúb-nubti, Kharnúb-shami.

This tree, a native of the countries bordering on the Mediterranean, attains a height of 20 to 30 feet. Its pods are known as carob or locust-beans. The seeds, which are small, hard, and reddish brown, are surrounded by a sweet mucilaginous pulp of agreeable flavour. Carob-pods are 6 inches to 1 foot in length, and about 1 inch broad.

COMPOSITION OF CAROB-PODS (C.).

	In 100 parts.	In 1 lb.
Water	14·6	2 oz. 147 grs.
Albuminoids	7·1	1 „ 60 „
Sugar (51·8 per cent.) and other Carbohydrates	67·9	10 „ 378 „
Oil	1·1	0 „ 77 „
Fibre	6·4	1 „ 10 „
Ash	2·9	0 „ 203 „

The nutrient-ratio is here about 1 : 8·5, and the nutrient-value 68. As sugar, pectose, gum, etc., occupy the place of starch in these pods, the starch-equivalent cannot be calculated in the ordinary way, for the sugar, etc., are of less nutrient worth than starch, containing for a given weight less carbon.

Fig. 35. Pigeon-pea (*Cajanus indicus*).

The carob-tree was introduced into India about the year 1840. Several attempts to acclimatise it have been subsequently made with varying measures of success. The failures which have occurred have generally been due to neglect of the plants during their young state, or to the unsuitable climate of the localities where they have been planted. The tree flourishes in a dry and poor soil, and requires such a climate as prevails in many parts of the Punjab, the North-West Provinces, Oudh and Rajputana. The tree fruits when it is 6 or 7 years old, and when in full bearing produces about 8 cwt. of pods per tree. In Cyprus, where the tree grows luxuriantly, several varieties are distinguished; the pods of the best kinds are less astringent than those of the wild sort. They are used occasionally in Southern Europe for human food, especially in times of scarcity, and are prized everywhere as horse and cattle fodder. For the latter purpose they are generally employed with beans or barley. The wood of the carob-tree is sound and hard.

THE INGA-BEAN.

Pithecolobium dulce, Benth.

Synonym—Inga dulcis (Willd.).

A middle-sized tree, belonging to the natural order Leguminoseæ, sub-order Mimoseæ, and tribe Acacieæ, with bipinnate glabrous leaves; a native of tropical America, but cultivated throughout India. The pod is 4 to 5 inches long, 6 to 8-seeded. The black seeds are half enveloped in a white pulpy edible aril.

COMPOSITION OF INGA-BEANS (C.).

	In 100 parts.	In 1 lb.
Water	13·5	1 oz. 70 grs.
Albuminoids	17·6	2 „ 357 „
Starch, etc.	41·4	6 „ 273 „
Fat	17·1	2 „ 322 „
Fibre	7·8	1 „ 108 „
Ash	2·6	0 „ 182 „

The nutrient-ratio is here 1 : 4·6, while the nutrient-value is 98. The indigestible fibre is rather high.

In the absence of chemical analyses of other Indian leguminous seeds, a mere mention of a few of the more important must suffice.

Psophocarpus palustris (Desv.).—A slender herb, a native of Africa, but cultivated in the tropics. It has a nearly rectangular pod, 2 to 4 inches long by ½ to ¾ inch broad, with 4 to 8 seeds. This genus belongs to the sub-tribe Euphaseoleæ.

Mucuna is a genus of Erythrineæ, a sub-tribe of Phaseoleæ. There are three species, M. pruriens (D.C.), M. capitata (W. & A.), and M. nivea (D.C.), unless, indeed, these be cultivated forms of but one species. They are annual herbs, having pods varying in length from 1 to 6 inches.

Tamarindus is a genus of Amherstieæ, a tribe of Cæsalpinieæ. The Tamarindus indica (L.) is a tree growing to a height of 20 to 30 feet. The seeds are sometimes eaten in times of scarcity. The pulp surrounding the seeds is sour, and is employed as a gentle laxative medicine; it also forms an ingredient in chutnees and curries. The moist tamarind pulp contains, on an average, about 5¼ per cent. of acid tartrate of potash, 6½ of tartaric acid, and 2¼ of citric acid.

It does not fall within the scope of the present work to enumerate and describe the various oily and saccharine fruits and the numerous fresh vegetables which are used in India as human food. The same statement must be made with regard to the condiments, spices, and flavourers which form such important ingredients in the popular dietaries of the country. Up to the present time we have at our disposal very few data concerning the composition of fresh Indian-grown fruits and vegetables. But we may safely affirm that their value as food mainly depends upon the fat or oil and sugar and other carbohydrates they contain; upon the presence of the organic acids, citric, tartaric, and malic; and upon their mineral ingredients, especially their potash and phosphoric acid. Other saline compounds, such as nitrates, also exist in these vegetable products; these are not without dietetic value. Taken as a group, these fresh fruits and vegetables, while adding to the combustible nutrients of the substantive articles of food, supply the mineral matters in which the latter are often, as is particularly the case with rice, notably deficient. They share also, with flavouring substances, the useful office of imparting variety and palatability to what would otherwise be a monotonous and insipid diet. This is a function which the physiologist, rather than the chemist, can more thoroughly appreciate.

Since the foregoing pages were in type I have been enabled, through the kindness of Mr. G. Watt, the Superintendent of the Indian Economic Court in the Indian and Colonial Exhibition, to secure a number of additional specimens for examination. Some time must elapse before they can all be analysed, but amongst them there are four examples of oleaginous seeds, of which I am able to give at once the percentage composition. It should, however, be premised that the quantities of oil set down amongst the constituents may include, in one or two cases, a little resin and colouring matter, and also that the albuminoids have been calculated from the total nitrogen found by analysis, and so may be somewhat exaggerated. Such specimens as required it were first of all freed from their indigestible coverings, the cleaned kernel being analysed in the state in which it is usually eaten or as prepared for cooking.

Shorea robusta (Gaertn.).

Hind.—Sál.

This is a large timber-tree of considerable importance, belonging to the Natural Order Dipterocarpeæ. It occurs in the sub-Himálayan tract from the Sutlej to Assam; and in the eastern districts of Central India, from the Ganges to the Godaveri, extending westward to the longitude of Mandla. The fruit is eaten by the Santáls in times of scarcity; the seeds are regularly consumed in a particular preparation with mahúa flowers.

COMPOSITION OF SÁL (SEEDS), C.

	In 100 parts.
Water	10·8
Albuminoids	8·0
Starch, etc.	62·7
Oil	14·8
Fibre	1·4
Ash	2·3

The nutrient-ratio, calculated from the above figures, is 1 : 12, and the nutrient-value 105. The latter number is probably above the truth, for it is likely that a not inconsiderable part of the nutrients in these hard seeds of unappetising appearance exists in an indigestible condition.

Pistacia vera (Linn.).

This small tree, which belongs to the Anacardiaceæ, is not a native of India; but large quantities of its fruits (pistachio nuts) are imported from Afghánistán into North-Western India, where they are consumed by all classes. The Afghan tree is almost certainly identical with that of the Mediterranean shores.

COMPOSITION OF PISTACHIO-KERNELS (C.).

	In 100 parts.
Water	5·9
Albuminoids	24·4
Starch, etc.	3·5
Oil	62·5
Fibre	1·3
Ash	2·4

The nutrient-ratio is here 1 : 6, and the nutrient-value 172. This Afghan sample contained 11 per cent. more oil than I found in European pistachio-kernels.

Buchanania latifolia (Roxb.).

Beng.—Chirongi. *Punj.*—Chirauli.

This tree belongs to the Anacardiaceæ. It is found in the sub-Himálayan tract from the Sutlej eastward, ascending to 3,000 feet. The kernels are much eaten as a substitute for almonds; but a fine oil, in large quantities, may be, and sometimes is, expressed from them.

COMPOSITION OF CHIRAULI-KERNELS (C.).

	In 100 parts.
Water	5·7
Albuminoids	27·9
Mucilage, etc.	2·7
Oil	58·6
Fibre	1·8
Ash	3·3

The nutrient-ratio is here 1 : 4·9, and the nutrient-value 165.

Pinus gerardiana (Wallich).

The neosia or edible pine is a moderate-sized conifer, found in the arid parts of the North-West and Punjab Himálaya, and extending into Afghánistán, at altitudes of 6,000 to 9,000 feet. The seeds are collected and stored for use. They form an important article of food in the regions where the tree occurs, and they are also brought into India by Afghan traders.

COMPOSITION OF NEOSIA-KERNELS (C.).

	In 100 parts.
Water	8·7
Albuminoids	13·6
Starch, etc.	22·5
Oil	51·3
Fibre	0·9
Ash	3·0

The nutrient-ratio is here 1 : 10·3, and the nutrient-value 154.

INDEX.

	PAGE
Albuminoids	2, 32, 119
Amaranths	104
Amarantus gangeticus	109
Amarantus paniculatus	107
Arachis hypogæa	127
Avena sativa	86
Bájra	56
Bamboo rice	102
Bambusa arundinacea	102
Bare-sustenance diet	20
Barley	99
Bean oil	8
Body-weight	19
Broom corn	85
Buchanania latifolia	177
Buckwheat	114
Buckwheat-allies	104
Bulrush millet	56
Cæsalpinieæ	122
Cajanus indicus	169
Canavalia ensiformis	144
Cane-sugar	5
Carbohydrates	4
Carob-beans	170
Catiang-beans	156
Celosia cristata	107
Ceratonia Siliqua	170
Cereals	31
Cereal-table	23
Chena millet	40
Chenopodium album	109
Chenopodium Quinoa	110
Chick-peas	128
Chick-peas, preparation of	120
Chirauli-kernels	177
Cicer arietinum	128
Classification of food	10
Coix lachryma	60
Common salt	15
Condiments	16
Conglutin	119
Cutin	7
Cyamopsis psoralioides	124
Daily rations	16, 25
Diet and work	13
Dolichos biflorus	162
Dolichos Lablab	161
Eleusine coracana	89
Fagopyrum esculentum	114
Fagopyrum tataricum	114
Fats	7
Fibre	6
Flesh-formers	3
Food-adjuncts	9, 11
Foot-tons	14, 19
Gluten	33
Gluten-casein	118
Glycerides	7

	PAGE
Glycine Soja	140
Gram	148
Gramineæ	36
Grasses	36
Great millet	80
Guar-beans	124
Guinea corn	80
Gundli	44
Hard-work diet	20, 29
Haricots	147
Heat-units	13
Hordeum vulgare	99
Horse gram	162
Indian corn	65
Inga-beans	173
Italian millet	55
Joár	80
Job's tears	60
Kángni millet	55
Koda millet	39
Lablab-beans	161
Lathyrus sativus	132
Legumin	119
Leguminosæ	122
Lens esculenta	136
Lentils	136
Lima-beans	155
Little millet	44
Locust-beans	170
Lupines	123
Lupinus albus	123
Mahúa flowers	6
Maize	65
Mannite	6
Millets	34
Mimoseæ	122
Mineral matter	2, 17, 34
Mixed diet	12
Moderate-work diet	20, 29
Moth-beans	152
Mucilage	6
Múng-beans	148

	PAGE
Neosia-kernels	177
Nitrogen compounds	3, 32, 118
Nutrients	9
Nutrient-ratio	14, 18, 39, 107, 123
Nutrient-value	14, 18, 39, 107, 123
Oats	86
Oil in rations	26, 27, 28
Oils	7, 34
Oryza sativa	66
Panicum colonum	50
Panicum frumentaceum	49
Panicum miliaceum	40
Panicum miliare	44
Papilionaceæ	122
Paspalum scrobiculatum	39
Pea nuts	127
Pearl-barley	101
Peas	135
Pease-oil	8
Pectose	6
Pennisetum typhoideum	56
Peptones	4
Phaseolus aconitifolius	152
Phaseolus lunatus	155
Phaseolus Mungo	148
Phaseolus vulgaris	147
Pigeon-peas	17, 120, 169
Pinus gerardiana	177
Pistacia vera	176
Pisum sativum	135
Pithecolobium dulce	173
Proteids	2
Pulse	118
Pulse, acreage of	121
Pulse, importance of	121
Pulse-table	24
Quinoa	110
Ragí	89
Rice	17, 66
Rice, dietetics of	75
Rice-oil	8
Rice, preparation of	73

	PAGE
Saccharum officinarum	76
Salt	15
Salts	2
Sánwa millet	49
Setaria italica	55
Shama millet	50
Shorea robusta	175
Sorghum saccharatum	85
Sorghum vulgare	80
Soy-beans	140
Standard diet	13, 15, 20
Starch	4, 31
Starch-equivalent	17, 23, 24, 36
Sugar	5, 31
Sugar-cane	76
Sword-beans	144
Uses of food	8
Vetches	131
Vetchlings	132
Vicia sativa	131
Vigna Catiang	156
Water	1
Water in Indian grains	1, 93
Wheat	34, 90
Wheat, acreage of	91
Wheat, character of	94, 96
Wheat, composition of	93
Wheat, mill-products of	95
Wheat, oil of	8, 34
Wheat, production of	92
Wheat, varieties of	98
Wheat, yield of	91
Zea Mays	65
Zein	33